A Project Manager's
Guide to Influence

Other Books by Colin Gautrey

Influential Leadership: A Leader's Guide to Getting Things Done

Advocates and Enemies: How to Build Practical Strategies to Influence Your Stakeholders

With Mike Phipps

21 Dirty Tricks at Work: How to Win the Game of Office Politics

With Dr. Gary Ranker and Mike Phipps

Political Dilemmas at Work: How to Maintain Your Integrity and Further Your Career.

eBooks

Positive Influence for Women

Building Reputations in Tough Organisations

Becoming More Influential

A PROJECT MANAGER'S GUIDE TO influence

COLIN GAUTREY
Foreword by Diane Dromgold

Copyright © 2015 Colin Gautrey Limited

First published in 2015 by
The Gautrey Group,
71-75 Shelton Street,
Covent Garden,
London,
WC2H 9JQ

Email: info@gautreygroup.com
Website: www.gautreygroup.com
Blog: www.learntoinfluence.com

Colin Gautrey asserts his moral right under the Copyright, Designs and Patent Act 1988 to be identified as the author of this work.

ISBN: 978-1-910470-09-1 (hardcover)
            978-1-910470-10-7 (paperback)
            978-1-910470-11-4 (ebook)

All rights reserved. No part of this publication may be reproduced, stored in a retrieval system, or transmitted in any form, or by any means, electronic, mechanical, photocopying, recording or otherwise without the prior permission of the publisher.

A catalogue record for this book is available from the British Library.

Typeset in Gandhi Serif by Ashton Designs.

Printed in the United Kingdom by Lightning Source UK Ltd.

Contents

Code of Conduct	11
Foreword	13
Author's Preface	17
How to Use This Book	19

CHAPTER 1

A Project Manager's Perspective	23
The Importance of Attitude	24
The Traditional Perspective	26
The Political Perspective	27
Power and Political Disturbance	33
The Relationship Perspective	37
Altering Your Perspective	39

CHAPTER 2

Project Managers under Siege	44
The Challenges	45
Exercise: Reflecting on Your Reality	49
An Alternative Reality	50
The Stakeholder Influence Process	52
Benefits of Application	55
Not Applicable Here	57
Remaining Realistic	61
Exploring Your Motivations	63

COFFEE BREAK

Taking Control	66

CHAPTER 3
Step 1: Focus   68
- Exercise: Establishing Your Priorities   69
- Deciding on Your Focus   70
- Turning Goals into Influencing Goals   73
- Becoming More Specific   79
- The Psychological Path of Goals   81

COFFEE BREAK
Lack of Authority   85

CHAPTER 4
Exploring Organisational Power and Influence   87
- Sources of Power   89
- What Power Means to You   94
- The Principles of Power   96
- Exploring the Principles   99
- Additional Concepts   100

COFFEE BREAK
Emotional Project Managers   107

CHAPTER 5
Step 2: Identify   109
- What is a Stakeholder?   110
- Identifying Stakeholders   111
- Stakeholder Categories   112
- Application to Your Goal   115
- Working with Groups   117

CHAPTER 6
Developing Political Insight   120
- Types of Agenda   122
- Exploring Political Agendas   123
- Distilling Political Insight   129
- Filling in the Gaps   130
- Making Comparisons   132

COFFEE BREAK
Translation Issues — 135

CHAPTER 7
Step 3: Analyse — 137
　The Stakeholder Map — 138
　The Relationship Position — 141
　The Agreement Position — 142
　Prioritising Your Stakeholders — 143
　Completing the Stakeholder Map — 146
　Drawing Conclusions — 148

COFFEE BREAK
Hopeless Projects — 152

CHAPTER 8
Understanding the Bigger Picture — 154
　Organisational Context — 156
　Developing Political Scenarios — 159
　Risks and Opportunities — 163

COFFEE BREAK
Diverging Objectives — 170

CHAPTER 9
Advocates, Critics, Players and Enemies — 172
　Engaging with Advocates — 173
　Engaging with Critics — 177
　Engaging with Players — 179
　Engaging with Enemies — 182

COFFEE BREAK
Project Governance — 186

CHAPTER 10
Step 4: Plan 188
What is a Strategy? 189
General Approach to Strategy Development 190
The Threads of Your Intelligence 192
Moving Stakeholder Positions 192
Exercise 196
Building your Plan 198

COFFEE BREAK
Steering Committees 200

CHAPTER 11
Building High Quality Relationships 202
Trust and Credibility 204
Communication and Influence 206
Problem Solving and Conflict Resolution 207
Assessing Your Relationships 209
Strengthening Trust and Credibility 211
Strengthening Communication and Influence 214
Strengthening Problem Solving and Conflict Resolution 220

COFFEE BREAK
Being a Stakeholder 225

CHAPTER 12
Step 5: Engage 227
Building a Compelling Vision 228
Create a Benefits Register 230
Tailoring Your Pitch 232
Adopting Alternative Tactics 233
Adapting Your Style 237
Managing the Politics 241
Preparation for Engagement 244

COFFEE BREAK
Remote Team Members 248

CHAPTER 13
Step 6: Maintain 250
    Increasing Motivation 251
    Motivation Exercise 252
    Exceptional Motivation 254
    Regular Review and Refreshment 255

CHAPTER 14
The Stakeholder Influence Process 261

CHAPTER 15
Project Hawaii 273

CHAPTER 16
Building Your Reputation 282
    How to Build a Reputation 284
    Project Teams 287
    Leading and Managing People 290
    The Benefits 292

RESOURCES
    Further Reading 295
    The Gautrey Group 297
    RNC Global Projects 299
    About Colin Gautrey 301

Code of Conduct

Everything you will read in this publication is based on the author's fervent wish that you will adopt an ethical approach to influence. This involves:

1. Always helping people to make balanced and informed decisions.
2. Ensuring pitches include the drawbacks as well as the benefits.
3. Being clear and open with people about your own interests.
4. Aiming for people wanting to do what you want them to do.
5. Never misleading people into doing something that you know will harm them.

Deviating from these ideals remains your choice and your responsibility. Gravitating towards these ideals will help you to prosper with a clear conscience.

Foreword

by Diane Dromgold

Projects and their management are changing. Not too far in the future, I predict over half of all money spent in organisations will be on projects. Projects that span departments, domains, territories, ideologies and political alliances.

No longer can we take a desired outcome, build a plan, budget, and then assume dedicated resources and ongoing blind commitment by the sponsor. No longer do we have the authority to demand and command resources to achieve the end. Today we're given an outcome in terms of the benefits with much more room for variation in delivery. We're told to 'get' the resources we need, knowing that means begging, borrowing and sometimes stealing. We can't make a plan and expect to stick to it, rather we have to lay out a path and gather different people to it at different times. We need to be able and ready to duck and weave and change and respond and be ready when naysayers criticize us.

I searched the world for books, seminars and resources that could help project managers interpret this world and succeed. I knew the missing link needed to be addressed in two ways, and one has to come before the other.

The first is to acknowledge that without authority over resources the only effective way is to achieve it through influence. Influence over stakeholders (positive and negative) and to be able to use that influence to guide activity and agendas through organisations. I even looked to political campaign management in case the answer lay there. My search uncovered Colin Gautrey and one of his earlier books, which I read in one sitting. I reached out and asked the initial question, "Could you write a book targeted at project managers?" As a former project manager himself, he didn't need much influencing, and here is the result.

This book adds significantly to the body of knowledge available for project managers. In fact, in my view, it supersedes a lot of the thought that's gone before. For the first time, we can read about the difference between project administration and project delivery and take comfort in the fact that project management, as we have known it, is now a subset of the bigger skill and capability requirements needed by project managers.

I thank Colin wholeheartedly for listening to my question, engaging with the subject, and responding with this book which explains, and provides opportunity to personally explore, becoming influential as a project manager.

The second way we can address the problem? Colin might shoot me if I give that away here, because the pages that follow will show you exactly how to become a project manager of the future.

Enjoy the book. I hope you find it as useful as I do. I'm giving copies to all our staff and clients as I don't mind RNC Global Projects staying at the head of the pack when it comes to enabling our people and yours.

Diane Dromgold

CEO, RNC Global Projects,

Sydney, Australia.

Author's Preface

At present, I am getting more enquiries for help with developing the influencing skills of project managers than for any other role. These requests rarely come directly from project managers. Usually, it is their leaders or the HR/Training people responsible for them who are alert to this training and development need.

I've been focusing entirely on helping people to become more influential for over ten years now, and I have to say, that the problems faced by project managers are not unique — but they are causing considerable pain. Frustrations and stress for individuals, organisations, and ultimately, for the people that they serve.

At the same time, the need for effective project management is increasing — rapidly. Effective project execution is vital in terms of realising the ambitious benefits expected by sponsors and clients. Indeed, failure to implement projects effectively is increasingly becoming a matter of survival for organisations. If they cannot land big projects fast and well, they will soon get left behind.

In the main, I believe that the major cause of this is the growing complexity of organisations and the environment in which they need to succeed. Today, many organisations span diverse countries, cultures and markets. To succeed, most organisations are dependent on a network of related organisations that work together to deliver what customers, clients and patients need. Consequently, projects have to be managed across many traditional boundaries in order to succeed for all parties.

So, we have arrived at an interesting time when the conventional toolkit of the project manager is declining in its effectiveness. Many of the approaches that have worked so well in the past are predicated on the notion that project managers have control, authority and almost perfect knowledge. These things are rarely possible (or even desirable) in the fragmented environment we face today.

Don't misunderstand me. I am not saying that the established ways are bad, it is just that today they are not enough. What is needed is to overlay a new set of approaches which can cope with the disparate demands of today.

The purpose of this book is to give you, a project manager, an opportunity to explore different approaches and ideas so that you can learn how they may be useful to you in the execution of your projects. There is nothing especially taxing in here, and many of the ideas may seem a little obvious at first sight. But when you bring them together and put them into action, the results are remarkable.

That they work is beyond doubt in my mind because I have seen them put into practice by ambitious and talented managers the world over. Many of these people are project managers, well qualified, experienced and successful. What they have found is that the simplicity of my approach, the structured nature of the frameworks, and the practical application, are especially suited to the world of project management.

However, I'll let you be the judge of that.

Author's Preface

If you grace me with your attention for a while, I will quickly introduce you to the concepts, ideas and models. You are busy I know, so my style will be succinct, no-nonsense, and I will do my best to make this material easily accessible. Many of the ideas here were first published in *Advocates and Enemies: How to Build Practical Strategies to Influence Your Stakeholders*. In this book, I have re-interpreted them from the perspective of project management, added additional content specific to your role as a project manager, and updated them to the latest thinking on influence.

As an author, I have to make some assumptions about you. This is necessary to develop a consistent approach to the style and content. In my mind, I am making the assumption that you:

o Are keen to do a good job.

o Don't have time to study lots of theory.

o Are intelligent and interested in developing your practice.

o Will invest some time in reflecting on how this all relates to your work.

o Are willing to make clear decisions about what you should do for the best.

You may not match all of these assumptions, and that is to be expected. Just be aware that as you read the book, this is what I have in mind.

How to Use This Book

My overriding objective with this book is to get you out of it again as quickly as possible. This may seem a little unusual for an author. My whole purpose in life is to influence people to take positive, high integrity action to become more influential. Therefore, as soon as you get the messages here, I'd much prefer you to just go and implement them and reap the rewards. Of course, you can come back and enrich your thinking and action as needed, but you really do need to take action on the basis of the ideas expressed here.

Consequently, each chapter provides suggestions to help you move more quickly through the material. Let me give you some now:

- If you are totally convinced that projects are political footballs and your role as a project manager is to make it happen in a rapidly changing social environment, go straight to Chapter 14. Have a good read and use the rest of the chapters to deepen your insight as necessary. Mind you, there is quite a strong case for skimming Chapter 1 first.

- On the other hand, if you are not quite convinced that project managers need to be exceptional influencers in order to succeed, make sure and have a good read of Chapter 1. Then skim the rest of the book before settling down to study Chapter 14. The other chapters can be read more closely later.

- Finally, if you're uncomfortable with the notion of becoming an influencer, or think that you shouldn't need to do all of that stuff, take a deep breath and read each chapter in order. Make sure you read Chapter 2 carefully, especially if you find Chapter 1 a little unsettling.

The most important concept here is action. These processes work. They are being used by project managers as well as marketing managers, IT specialists, lawyers and executives at every level in some of the biggest corporations in the world. If they can work for them, they can work for you too.

Consequently, you can use the processes to work directly on projects you are managing, on things that need to change around your role or even, on moving forward your career. You do not need to apply these ideas just to your work as a project manager.

One more thing before you get stuck in. Often, the biggest obstacle to becoming an effective influencer is remaining mindful of what you've learned, or have always known you need to do. With such a busy life it is difficult at the best of times remembering to do what you know you should be doing, particularly when you have lots of people putting you under pressure for all manner of things.

So, you might like to find a way of keeping this present in your mind, a way of being reminded that it's a smart idea to do it. One way of doing this, which I would encourage you to do, is to subscribe to my **Influence Blog**. The purpose of this blog is to give you regular ideas about how to become more influential. When it drops into your inbox, it will be a good reminder for you, and yet another way of improving your practice while also reminding you to get your stakeholder plan out again.

You can find the Influence Blog at:

> www.learntoinfluence.com

Blatant, but well-intentioned self-promotion over, time to get moving.

Colin Gautrey

London, December 2014

CHAPTER 1

A Project Manager's Perspective

This book is all about taking action, and I want to hit the ground running.

Most problems and challenges in life are a matter of perspective. Adopting alternative perspectives enables you to see the problem differently. It doesn't mean that one viewpoint is right or wrong. It is simply a technique that can give you new insights and potentially, a solution that will work.

So, in this first chapter I want to challenge you to think about project management from a few different perspectives.

This chapter will help you to:

- Understand how attitudes impact action — yours and others.

- Explore your attitudes and beliefs towards projects.

- Analyse the politics around your project.
- Determine the implications for your approach to project management.
- Reflect on your view of what project management is all about.

If you're in a hurry:

- You need to judge the time to rush and the time to pause. My strong advice is to pause and read all of this chapter, especially if you are a little uneasy about the political nature of projects.

The Importance of Attitude

Two people faced with exactly the same situation will, if they hold different attitudes, see things differently. They will also make different choices as a result of what they see. The ripple effect of these choices can create a huge difference in the performance and results that they get. Imagine the following scenario:

Peter had allocated personnel to two projects run by Ann and Ajay. Both projects were heavily reliant on Peter's people. Due to an unexpected crisis, Peter had to reallocate his resources temporarily and sent an email to both project managers explaining the situation, apologising for any inconvenience. He also promised to send the people back as soon as he could.

Ann had a negative attitude. She believed that this was typical and illustrated a lack of commitment to what she was trying to deliver. It confirmed her suspicion that Peter didn't want her to succeed. After all, it had taken lot of pressure to get the people in the first place. Consequently, she replied immediately wanting to know exactly why he had done this and demanding to know when the resource would be returned.

Ajay had a positive attitude. He believed that Peter would only do this if he absolutely had to. It was difficult to get the people in the first place, but with a bit of negotiation they had managed to secure the necessary resource. If Peter wanted them back, it must be serious. Consequently, he adopted a different approach to Ann. His response to Peter recognised the urgency, promised to help his people to refocus quickly so they could deal with the crisis. He also asked if there was anything else he could do to help.

Although the problem presented to Ann and Ajay was the same, their response is very different. It is unlikely that Peter's behaviour would be massively affected in the short term; he will certainly be feeling very differently towards each manager — wouldn't you? Should he be able to begin reallocating the resource back to the projects, who do you think he will favour?

The impact of their attitude would also affect others around the situation. Ann is likely to try to cling on to the resource, attempting to pull them back and talk negatively to anyone who will listen about how difficult Peter is being. Ajay on the other hand is more likely to continue voicing his support, helping make things as easy as possible for Peter and explaining to other colleagues that they need to remember the bigger picture.

The impact of these attitudes is far reaching and often unseen until they present a problem. If you had a choice, whose project team would you prefer to work on? How easy would it be for each to recruit team members? How long would team members stay? Which project manager do you think is enjoying their work the most?

Attitudes are built on many beliefs. Beliefs work at a subconscious level and affect how you interpret events. Attitudes are what make you look at the world in a particular way, your perspective. Because they are generally outside of awareness, attitudes are often automatic and seemingly beyond control. Hence, people tend to adopt a perspective automatically almost out of habit.

But it doesn't have to be that way. To adopt a different perspective all you have to do is pretend that the beliefs and attitudes necessary for it to be true are true. That doesn't involve you denying what you currently believe. All it needs is for you to suspend judgement and engage your imagination a little.

To begin, let's start with the perspective most project managers adopt.

The Traditional Perspective

Project management is a well-established discipline, and consequently has some fairly stable beliefs, processes, methodologies and so forth. I expect that the vast majority of project managers subscribe to this traditional perspective when they are considering their projects. Before I share my thoughts on what this is, I'd like you to take a little time and consider your answers to the following three questions:

- When you think about the topic of projects, what thoughts and ideas come to mind?
- What are projects designed to do?
- What are the three most important things to remember when it comes to projects?

There are no right or wrong answers here, just your views based on your accumulated experience. You might have thought of some of these things:

- The purpose of projects is to organise activity towards a specified goal.

- Performance against clear QDC standards is paramount.
- Projects should draw resources from a wide section of interested parties.
- Strict controls and governance need to be adhered to.
- Projects are impossible to control due to part-time resources.
- Projects represent a logical way of getting things done.
- Resources allocated to projects are often those not wanted anywhere else.
- Everything needs to be meticulously planned.
- Use of project management is critical for all bar the smallest of projects.

These are just a small sample of what you may have come up with. Before you move on, take a time-out and draw some conclusions about your current perception of projects:

1. What is your definition of a project?
2. What three beliefs are the most important when it comes to managing projects?
3. How would you summarise your attitude towards projects?

Now, let me stir it up a little.

The Political Perspective

I'd like to propose that the purpose of a project is to adjust the distribution of power within a social group, organisation or system (such as a supply chain), usually under the guise of legitimate activity. Even when this is not the deliberate purpose of a project, the outcome nearly always results in a re-distribution of power. This may appear a little radical or off-the-wall, and I don't expect you to buy into this right away, just keep an open mind for the next few pages. In a later chapter, I'll go into the subject of power at length.

From this perspective, you might have projects that are:

o   Initiated by those who wish to gain more power and/or protect their position.

o   Supported by those who are going to retain or build more power.

o   Resisted by those who expect to lose power.

o   Subtly influenced by powerful stakeholders.

o   Led by people seeking personal gain, especially by way of new permanent positions.

o   Resourced in a political manner.

o   Of somewhat questionable benefit.

o   Political footballs.

Does any of this sound familiar? I'm sure it does, and that's because these things are extremely common, and all bear witness to the reality that the vast majority of projects impact the power structure or dynamics of the host organisation. It also means that projects can have a massive impact on the hopes and fears of people working there.

Here are a few examples of projects that support the reality of the political perspective.

Company Restructure Project

Sally was responsible for delivering an organisational restructure project. She needed to manage the resources to analyse the current structure and the problems they were facing. This involved her team members executing a wide-ranging research initiative which involved staff, clients and suppliers. Her sponsor was the Customer Services Director.

Eventually, the team drew their conclusions and were ready to present a new structure for approval by the board. It addressed many of the problems the organisation was facing and also introduced some additional benefits that could lead to faster product development and enhanced customer service. On the face of it, how could anybody object?

In the coaching session prior to their presentation, I was interested to know who Sally considered to be the most powerful people in the organisation and what impact her proposals would have on them. Without hesitation, she told me that the Sales Director was by far the most powerful person. He was responsible for the big ticket relationships that generated almost two-thirds of annual turnover. If he left, so would the business, and this gave him enormous positional power.

Unfortunately, the solution they had arrived at would result in the sales channel being divided into three. Put another way, her proposal was to divide his power by three. As she started to explore the political setting for her project she realised that if it were successful, her sponsor, the Customer Services Director, would then be much more powerful because of the break-up of a competitor's territory. Not because he would himself grow in power, but because a key power source would be broken up.

At no stage had this been addressed. The proposal was heading to a decision-making body where the Sales Director was hugely influential. Little surprise that Sally did not succeed in getting the go-ahead to restructure the organisation. It didn't die straight away, but it did in the end. Basically, it was a costly and perhaps naïve, attempt on the part of the Customer Services Director to gain power.

Interestingly, a few years later, I heard that the restructure had gone ahead without a costly project, and with no opposition. They had made the Sales Director the CEO.

Management Information System Project

Executives need information in order to make decisions. The more up-to-date it is, the better. However, for many executives, this is a distant dream. Despite the technologies available, they still have to wait for the end of month/quarter figures to become available.

Marcus landed the project to implement a new Management Information System and was excited that he had the full backing of the executives. They wanted this to happen and expected it to be a reasonably simple thing to do. So did Marcus.

Yet, when he went to work, he met with silent but persistent resistance. It seemed that the only people who wanted this to happen was the executive team. He wasn't surprised to find that the finance team was reluctant. A key power source for them was the ability to generate the numbers. Their work was meticulous and took concentration. Therefore, you disturbed them at your peril. The numbers arrived when they were ready, and everyone else, including the executives, had to wait. An up-to-the-minute MIS would remove that source of power, and they had little else to fall back on.

What did surprise Marcus though was the resistance from the other business teams, especially the operational teams. They were throwing up all manner of problems and reasons why it would not be possible to contribute in a meaningful way. Although it was not admitted, he suspected that the real cause was their desire to avoid closer inspection. In the current set-up, they had plenty of time to resolve temporary problems in their processes. Once the MIS was available, the executives might be jumping up and down as soon as a process wobbled. In many ways, this was reasonable. Instead of spending time answering executive questions, they could be focusing on fixing whatever was wrong.

All around, Marcus was noticing that information flows and controls would be changing as a result of the implementation. Ownership of information would be changing (position power) and that was unsettling everyone. What originally started off as a purely technical implementation quickly became one of managing and negotiating at a relationship level. In fact, one of the results was to pay closer attention to the executive decision-making process once they were in receipt of the available data.

To put it bluntly, with the best of intents, the purpose of this project was to shift positional power to the executive team from the accountants and the operations people. What had not been anticipated was the reaction of those who would be losing power.

New Business Division Project

This is actually one that I was the project manager for many years ago. I was coming from one of five existing divisions onto a team established to build a new business. The initiative was the outcome of a lengthy investigation by external consultants into new business opportunities our company was capable of exploiting. Being somewhat naïve, I approached this from the traditional perspective, needing to quickly organise the plans, resources and so on. Luckily, my intelligence quickly caught on to what was really going on.

The first clue was that, to me, the business case didn't stack up. Although the numbers were compelling, the business model just didn't make sense. My reservations were dismissed by the guy leading the initiative (let's call him Bob) and also the lead consultant (Marcel). They were both very intelligent so maybe I'd got it wrong.

The next clue came from my Managing Director, one of the existing divisional heads. "Colin, the reason I've put you forward for this is to protect my investment and keep me informed of what is going on." Who was I to argue? He'd been my boss for several years, and I had a huge amount of respect for him.

Two further things happened which brought the political perspective into focus. Firstly, Bob began to confide in me that when the project landed I was well placed to be appointed to a senior position in the business. His expectation was to create this as a new division alongside the existing ones (very different from what my boss told me would happen). Added to this, I had also overheard Marcel, talking excitedly to a friend about how important this project was to his partnership prospects.

In a nutshell, Bob was using the project to become a divisional head and Marcel to gain his partnership within the consultancy. Much of the projected £3.8m build cost would be heading in their direction along with a large amount of the annual expenses. That the profit was a pipe dream appeared irrelevant to them. Maybe they genuinely believed it would arrive. Personally I suspect that they did and that their personal ambition had removed their ability to be objective about the situation.

By taking the political perspective, I was able to see what needed to be done. I did my duty to the organisation. Bob and Marcel didn't even see it coming.

If you adopt the political perspective for projects, you will quickly begin to see them in a new light. It doesn't exclude the validity of a more traditional perspective, it simply allows you to think about it in different ways, generate alternative insights and plan action which may not otherwise have been obvious. It will also help you to learn how to deal with the challenges you face far more effectively. I'll talk more about how you can do this at the end of the chapter.

Power and Political Disturbance

Aside from the direct impact of projects on the distribution of power, power can also influence the smooth running of projects in a wide variety of ways. This is because no project exists in isolation. Around it are all manner of other projects, initiatives, processes, and yes, personal agendas. Gaining an acceptance that these things are happening will help you to become more objective about the problems and issues they create for you. It will also help you to learn how to mitigate the risks they pose.

Here are some of the more common ways that power affects projects:

Resource Allocation

If powerful people want to use their resources somewhere else, they will. If your project is not an important thing on their agenda, they will be reluctant to give you their best people, or may delay supplying the resources. It might even be that the project benefits them but grants greater power to one of their adversaries.

This is a good illustration of the principle of supply and demand. The allocators have the power to redirect supply, which you have a demand for. It doesn't matter why they want to do that, if they do, it is going to affect you. It also doesn't matter about the rationale for your project, if they feel they have the power to divert resources, they will.

Given the challenge of insufficient resources in project management today, this becomes a critical priority for the project manager and sponsor. Making sure that there are no political barriers to gaining (and keeping) the resources they need is vital. So too is ensuring that they appear on time and are proactively managed. Reacting after the problem arises is already too late.

Part-Time Team Members

It is rare these days for team members to be allocated on a full-time basis. More likely is that they will retain their day job. In most cases, they will come under pressure from their line manager to continue performing their normal tasks. On the surface, the line manager will be supportive of the project and play the role of being a good corporate citizen. Unless the project provides them with sufficient benefit, it will not be long before they are putting pressure on their team member to deprioritise their contribution to the project.

Line managers and team members will be weighing up the consequences and alternatives as they realise they don't have enough time to do everything (this is a great illustration of some of the principles of power I will introduce you to in Chapter 4).

Thus, team members are placed in an invidious position, stuck between their desire to do a good job for your project, and doing what their boss wants. Remember, it is unlikely they will be rewarded financially for their work on your project, so it is likely that the influence of their day job will prove irresistible. Empathising with this almost inevitable struggle for commitment will help you to build stronger relationships with your team members.

Divided Loyalties

Team members are often drawn from a spread of different departments or divisions. Team members usually return to their department after the project has completed. While they may be committed to your project, they will also have one eye on repatriation back to their original position. This will influence the decisions they take, especially if these may bring them into conflict with their original colleagues and managers.

Again, it is wise to recognise this potential so that you can deal with it more effectively. Although it will be hard to remove the problem altogether, you can factor it into the decisions you are asking individuals to back, and the tasks or meetings you allocate to different people.

This is a good reason why it is worth investing time in really getting to know the people on your team so that you can understand and manage this risk.

Executive Time and Attention

In all large organisations, there will be a large portfolio of projects and initiatives in flight. The powerful executives will be focusing on those which offer them the most potential to achieve their objectives (or rather, which will give them the most power). Consequently, the project you are managing will have political value at different levels for different executives.

Executives will compete with each other to ensure that their favourite projects are focused on, decisions made and progress advanced. When this means they need the input and approval of their fellows, they will influence it to the top of the agenda of any meeting.

If your project doesn't have the right level of political interest with the right people, you could find yourself waiting patiently outside of the board room for your time to present, and going back to your desk without ever seeing a board member. Even if it does get discussed, it may not get the attention it needs and deserves.

As a project manager, managing a project in the political backwater is hard work.

Political Value

You can learn a great deal about the relative political value placed on a project by noticing the senior people who are taking an interest. Important projects will have powerful sponsors. Indeed, they may even seem to have many sponsors in addition to the one nominated.

It is also important to note the quality of the appointments. Mission critical (powerful) projects will have heavyweight resource allocated. They will also be well positioned within the political structure. The best talent will be offered and/or they will be pitching to get involved.

In my early days as a project manager, I used to hate the imposition of strict project definition documents and steering committees. Governance procedures were an irritation at best. Now, with the benefit of experience, I know the job these things do. I also know that it can indicate how valuable a project is to the organisation.

Projects with high value generally get the attention they need and the resources, to succeed. Those with lower value will be persistently troubled and have a hard path to follow.

Rather than continue to cite examples for you, take a few moments to reflect on your experience:

○ How have you noticed power affecting your projects?

○ What is the political rationale for the last project you worked on?

○ Can you think of examples where projects have failed because of the politics?

Before you start to think that I am attempting to turn you into a political crusader or a Machiavellian schemer, I'd like to introduce you to another perspective for project management.

The Relationship Perspective

In reality, the political perspective may not be practical for you as a project manager. It will be extremely useful in helping you to understand what is actually going on around your project, but will it result in you becoming politically active? Probably not, for a number of reasons:

- You have been given the job of project manager. The organisation is expecting you to organise and deliver.

- Most of the political dynamics surrounding projects are above the level of most project managers in terms of position and arguably, capability.

- Powerful people will be resistant to you becoming an actor and perhaps making life more complicated and risky for them.

- While you may think you have a good political understanding, as soon as you get involved, you will quickly discover there is so much more that you don't know.

This is not me being disrespectful of your capability, nor wanting to dampen your enthusiasm and ambition. Instead, I want to offer you a bit of realism and also position this appropriate to the majority of project managers. The more experienced and senior you are, the more likely it will be that you will (and should) engage politically on behalf of your projects. For most though, you simply need to open your eyes and begin learning fast about the political reality of your project, and then adopt a relationship perspective to your job as a project manager.

The relationship perspective suggests that the project manager's role is to become a manager of stakeholder relations. The main job is to achieve and maintain consistent agreement and commitment to achieve a mutually desired outcome. Absolutely, you have to get a result. That is what you are engaged to do. That it may change along the way is to be expected. The relationship perspective will help you to remain in that process of inevitable change; it will also be possible for you to facilitate and control it too.

Considering your project with this perspective would incline you towards activities such as:

- Building strong relationships with key stakeholders.
- Creating high levels of trust so that they open up to you.
- Really listening to what they are saying and meaning.
- Helping stakeholders to work together, settle their differences and reach genuine agreement.

In effect, it will put you in a neutral position of trust with all stakeholders and evolve your role into that of a facilitator who is intimately connected with the more technical/resource/task implications of the evolving decision process. In this position you are likely to be able to influence the life of your project, your team members and alleviate many of the problems facing project managers today.

Altering Your Perspective

The three perspectives presented above can be summarised as:

- **The Traditional Perspective**: Where projects are intended to coordinate tasks and activities towards a predefined and agreed outcome.
- **The Political Perspective**: Where projects are intended to change the power structures of a group or organisation.
- **The Relationship Perspective**: Where the role of the project is to facilitate the delivery of a mutually desired and agreed outcome.

The choice is yours. In fact, choose them all in turn. You can use each perspective to bring new insight to your work and inspire alternative actions you can take. Adopting a given perspective doesn't mean that you have to agree that it is right, merely accept that it is an alternative way of looking at your project.

An Exercise in Perspectives

Consider a current project you are working on and adopt the political perspective:

- Whose power will increase if your project succeeds?
- Whose power will diminish if your project succeeds?
- Which stakeholders are disagreeing with each other? Why might this be?
- What political motivations could be working behind the scenes?
- What political motivations or allegiances may your team members have?
- What evidence is there for your thinking?

Don't let this make you paranoid. This is an attempt to understand the softer side of your project and the political drivers which may be influencing the way project decisions are being made.

Now, adopting the relationship perspective:

- What do you need to do in order to confirm your suspicions?
- How is the political backdrop affecting what is happening within the project?

- Where are the major political disagreements between your stakeholders?
- What action can you take to get closer to the parties concerned?
- How could you facilitate a resolution?
- What action can you take to protect your project?

A Reality Check

The last question is a little problematic for me. Personally, I am not sure that it is right to protect your project from its environment. From a traditional perspective, you might consider that once a project has been formulated, that's what it needs to deliver. Well yes, and no.

Toeing a hard line about the agreed deliverables is to defy the reality that the host organisation is trying to solve the challenges it faces in order to become more successful. At a given moment in time, a snapshot was taken, and the consensus was that it needed to change in a particular way. Whether it was eminently sensible or politically convenient doesn't matter. The point is that it was decided to deliver a particular objective. Assuming that it was right at that time doesn't mean that it will always be right.

Within the organisation, power will be ebbing and flowing. Battles will be waged, wars won, and peace made. The competition for power (like it or not) is raging in most organisations, often just below the surface. On the top of this comes the myriad of projects and initiatives supported and challenged by the personal motives foremost in the mind of many the contenders.

Yet this is not necessarily a bad thing. As I like to point out, power is responsible for all that is good too. What makes it good or bad is the intent behind the moves, the lengths the players will go to, and the depths they will stoop to, in order to realise their agenda.

In fast-moving organisations and environments, especially competitive ones, change is happening all the time. In many ways, it is almost inconceivable that any project will remain the right answer to their problems. Project managers who hold on to their deliverables like grim death are those who will face the most significant challenges. Those who settle into a more dynamic world and drive for delivering what the organisation needs will be those who thrive, and have the most fun.

That doesn't mean putting the agreed deliverables to one side or ignoring them. Instead, it means progressively evolving them in a carefully managed way. By attending to your job as a facilitator of change, you are more likely to deliver value to your host and have a great deal of fun in the process.

Key Points

- Attitudes and perspectives have a dramatic effect on thought and action.

- The purpose of adopting a different perspective is to stimulate new thinking.

- Beliefs are harder to change than perspectives. You can simply decide to adopt a particular perspective to help think things through.

- Virtually all projects result in a change to the power structures of groups/organisations.

- If you threaten someone's power, expect resistance at least.

- If your project boosts someone's power, expect support at worst.

Suggested Actions

- If this chapter has challenged your notion of what project management is about, take a few days out to reflect and discuss with a few colleagues.

- For each project you are working on, set aside some dedicated time to consider them from a political perspective. Keep an open mind and see where it takes you.

- Review your thinking with your line manager, mentor, coach or sponsor and see how they react.

- If you're not quite sure about what I've shared here and its relevance to your work, take your time reading the next chapter which looks at the challenges project managers are facing.

CHAPTER 2

Project Managers under Siege

If the last chapter was unsettling to you as a project manager, this one will help you take a more relaxed look at what is going on around you and what it means for your role.

This chapter will help you to:

- Review and reflect on the major challenges facing project managers today.

- Pause and explore the challenges you are facing right now.

- Gain a high-level understanding of the overall process proposed in this book.

- Learn how this process is working for others.

- Consider the benefits you will gain if you can overcome these challenges.

- Challenge you to step-up and take responsibility for making things happen.

If you're in a hurry:

- Just skip this one — provided you are now convinced that it is critically important for project managers to become effective influencers.
- Otherwise, skim and focus on the section, *Exploring Your Motivations.*
- Oh, if you're one of those heavy-hitting contract project managers, make sure and read the *Not Applicable Here* section at the end.

That said, this is an important foundation chapter, and it is well worth reading properly, even if you don't do the exercises. If you have other project managers reporting to you, it is well worth reading fully.

The Challenges

This is not a book about project management. This is a book about helping you as a project manager to become more influential. Personally, I believe that with strong relationship management skills (or rather practice) and greater attention to the job of influencing, most of the challenges faced by project managers can be alleviated or removed altogether.

Before writing this book, I wanted to update my knowledge on the topic (yes, I used to be a project/programme manager in a previous life). After a great deal of reading and talking, it became apparent that the key challenges facing project managers today are:

- Undefined or unclear vision, goals and objectives. Even when these are agreed and written down, diverse stakeholders may hold various interpretations of what has been agreed. Sadly, these usually only come to light after implementation has begun.

- Unrealistic expectations and demands. In the main, this relates to excessively tight deadlines. Conspiring with this is the fantastical belief that resources are superhuman and can hit the deadlines without any loss of quality or sleep.

- Misaligned goals between projects and the organisation. Most projects make perfect sense in their own right. When they have to fit into the bigger picture, problems begin to emerge, especially when different executive sponsors are involved.

- Conflicting priorities. In addition to the obvious inter-project rivalries, project managers increasingly have to compete with project team members' day jobs.

- Changing priorities of the organisation. Given that most organisations inhabit a chaotic environment, it should not come as a surprise that the demands of the organisation are moving faster than the project manager's ability to implement.

- Poorly managed issues, changes and risks. In the light of the rapid changes around projects, little wonder that staying on top of all these things is challenging.

- Lack of accountability/responsibility. From the perspective of project managers, this is usually levelled at sponsors, key stakeholders and users. However, those people often claim that project managers are failing to take appropriate responsibility.

- Poor communication. Usually, at every level and in every direction, communication could be better. As change around a project increases, so too are the harmful consequences of ineffective communication.

- Lack of buy-in, support and engagement. This can come from many directions, but especially frustrating for project managers is the apparent disinterest among key people like sponsors and users. Just getting an email response is sometimes all but impossible.

- Inadequate skills. In addition to the team members and other project resources, many project managers (especially in smaller projects) feel they lack the skills required to do the job, particularly when it comes to engaging with senior stakeholders.

These seem to apply to a very broad spectrum of projects, from the very small and informal right up to the huge global and multi-organisational projects. With the most complex of projects, additional challenges are being thrown up:

- Maintaining alignment with inter-dependent changes/projects. In large organisations, there are an awful lot of things changing all the time. Stability is unthinkable. To believe it feasible to be able to manage, control and co-ordinate all of these is unrealistic. To be honest, in many places, it is difficult to even be able to uncover all of the inter-dependencies until someone shouts because the project has tripped them up.

- Fostering team spirit among dispersed/matrixed project teams. The geographical spread of project team members makes it unusual to be able to bring people together in one physical location. On big projects, this may only happen a couple of times a year at best. Given the cycle time of many projects, building personal relationships between team members is unlikely to happen. Even between the project manager and each team member this is ambitious.

◦ Understanding the political environment in large organisational structures. Many project managers arrive in their role from more technically orientated roles. This means they are often unprepared and unwilling for the political world they have landed in.

◦ Influencing resources without formal authority. This is really an extension of a challenge already mentioned. In larger and more complex projects, the number of people reporting to the project manager may be zero. Consequently, their ability to influence hierarchically is very limited.

◦ Changing sponsorship/governance structures, processes and people. I've heard it from many sources that the setting for many projects is so complex that the project manager can easily lose track of who their key stakeholders are, let alone what their stakeholder expectations are. Even when the stakeholders do remain the same, their requirements can change like the wind.

I don't intend to delve deeper into the challenges, but I do want you to. What I've laid out above is based on my research; it is not intended to be exhaustive. These may vary from your own experiences. With all personal development, it is important to increase awareness of your own position. Then you will be better placed to interpret the ideas covered into this book, and how they will be able to resolve the challenges you are facing.

Exercise: Reflecting on Your Reality

The purpose of this exercise is to help you to review where you are now. There is no need to deploy your Root Cause Analysis talents right now, just invest some time here in thinking through what you are experiencing at the moment as a project manager.

Which of the challenges cited above are you experiencing at the moment?

What challenges are missing from the list that are having a big impact on you right now?

Of the challenges you are facing now, identify 3 or 4 which are causing you the most trouble.

For each one, think of a specific example of an incident that has arisen recently, then answer the following questions (for each one):

- What was the problem?
- Who was involved?
- What did they do?
- Why did they do what they did?
- What were you doing? Why?
- How did it end?
- What impact did this have on you and your work?

Do you notice any themes or trends in the challenges you are facing?

A little later, I will ask you to dig deeper into the consequences of these challenges. Before I do, I'd like to introduce you to an overall process that you'll explore in this book — the Stakeholder Influence Process. When applied thoroughly, it will enable you to make significant progress on all of the challenges you are facing.

To begin, here is an example of a client of mine who has successfully integrated the process into the way he works — to the immense benefit of himself and the organisation he works for.

An Alternative Reality

When I sat down to coffee with Mark, he opened his notebook and showed me a collection of diagrams at the back — each not much more than a few lines and scribbled names. Here was the evidence that Mark was still applying the Stakeholder Influence Process I had introduced him to five years earlier. It had become an integral part of his *modus operandi*.

Originally, I was asked to work with Mark to help him develop his approach to gaining stakeholder buy-in. As an experienced (and accomplished) project manager he had evolved his own way of engaging people. However, in a complex and fast-moving global business, it had been recognised that this was not as effective as it could be. So we designed a programme of six, one-hour, remote coaching sessions.

When we started to work together, I insisted that we agreed on a specific project and goal that we could use as a focus for our discussions — one that he and his organisation would gain great benefit from when realised. His top priority right then was implementing a pan-Asian procurement process. With this as our focus, we got to work.

Mark was struggling with many of the challenges mentioned earlier. Not only was he working in a highly complex environment with multiple agendas running, he was also newly appointed to his role. In one of our sessions, I showed him how to take a practical and strategic approach to engaging with his stakeholders. We mapped out his stakeholders, analysed them, and discussed their agendas, issues and his relationship with them.

A key problem he was facing was the resistance from various country heads who were quite happy with their own suppliers and were worried about the potential risks (and costs) of switching to suppliers they had not dealt with before. Although not stated by these stakeholders, Mark had a strong suspicion that their perceived loss of decision-making authority on procurement within their countries was also a factor.

Once he had explored the political ramifications of what he needed to agree to, Mark quickly worked out a strategy for gaining their buy-in. Simple actions that had not occurred to him before. None of them involved the deployment of new skills. Nor did they require him to go into battle, chase conflict or force people to agree with him. All that was needed was an effective way of thinking it through and tapping into his existing knowledge and skill. All this happened in a single hour-long telephone call.

Within six months of starting work with Mark, he had succeeded in getting his procurement process signed off and implemented. This was projected to yield first-year savings of over $28 million. He cited the stakeholder management approach as being critical to gaining the support he needed, overcoming strong opposition, and helping him to land the result. A $28 million result.

Of course, there was much more to achieving the result than a little bit of stakeholder management. He had to deploy high levels of intelligence, skill, logic, persuasion, etc. However, what it did do was ensure that he got focused on the people who mattered most, was able to think through his approach and realise the critical arguments which he needed to win in order to get his result. The evidence at the back of his notebook shows that it has remained an approach central to all of his important work since then — and it took just an hour to learn.

Mark is just one of many examples I could give of how people at all levels have benefited from the ideas, concepts and techniques contained in this book. Over the last ten years, I have worked with thousands of people at all levels and in many countries. Almost without exception, they have found that this simple approach helps them to focus on the important considerations affecting their stakeholders.

Here is an overview of the high-level process which I showed Mark how to use, and which forms the foundation of everything within this book.

The Stakeholder Influence Process

The Stakeholder Influence Process is a sequence of steps that will help you work out what you need to do to achieve your goals when they are reliant on the agreement of other powerful people. It provides a simple framework to think through the situation you wish to influence, your goal or the project you are managing. It will help you to figure out who is on-side and who may be out to get you. It will enable you to develop a clear strategy to influence the achievement of your goal.

This strategy will probably involve engaging with people you may have overlooked before. Before starting the process, you may have considered them to be minor players or not even interested in what you are doing. Additionally, the Stakeholder Influence Process may uncover opportunities to collaborate for mutual benefit, or perhaps, simply bring their attention to all that they could gain from your work. If they are powerful people in your organisation, this alone could help you to reach your goal.

Don't let the simplicity distract you. Within each step are some vital principles which add significantly to the potential of the process. Principles established through years of practical application in a wide variety of situations. These will be explored in later chapters.

The Stakeholder Influence Process

1 Focus
2 Identify
3 Analyse
4 Plan
5 Engage
6 Maintain

Step 1 — **Focus**: Assess your priorities and focus your Influencing Goal.
Step 2 — **Identify**: Work out which stakeholders can have the biggest impact.
Step 3 — **Analyse**: Map the position of each stakeholder.
Step 4 — **Plan**: Decide your strategy for increasing buy-in.
Step 5 — **Engage**: Adapt your approach to influence your stakeholders.
Step 6 — **Maintain**: Keep the momentum going with regular reviews.

Right now, I want to stress that this process is iterative. You can speed through the process once and get new actions jumping out all over the place. Then you can come back and do it again and go a little deeper. Each time you do this, there will be more things you can think about, more actions you can develop, and more progress you can make.

The Stakeholder Influence Process is not about complex theories of human behaviour. Instead, it focuses on simple questions like:

- What does each stakeholder want to achieve?
- What is the quality of your relationship like?
- To what extent is their agenda affected by what you are doing?
- What's going on around them?
- What are you going to do with them?

Exploring the answers to these questions and more, around a simple framework, will quickly help you to discover the key moves you need to make. Usually, these moves can be achieved with your current skills. Most people don't need to develop new skills in order to become more influential. Often it is simply a question of applying existing skills in different ways or with a different focus. Sometimes, you just need to do what you have been avoiding doing for far too long, and the Stakeholder Influence Process will help you to recognise the stumbling block and find ways to get moving again.

To be totally frank, the key here is in gaining the awareness of what you need to do. There is no deep psychology, just a simple process which demands crisp answers to basic questions. And it gets results — time and time again.

And to continue being frank, this process is not suitable for everyone. To get maximum value out of this approach, you need to be able to tick most of the following:

- You recognise that influence is a key part of your job.
- You need to gain the support of a variety of different people.
- Your goal has natural opposition.
- Many different opinions exist about what the right answer is.
- You want to become more influential.
- You are very busy at work.
- You are ready and willing to give it a go.

The most important item on the list above is the last one. Without this, you are probably going nowhere. I cannot think of a single individual who has failed to benefit after putting pen to paper and following the process. And it doesn't take much effort or time to make real progress.

Benefits of Application

Over the years, I have seen people achieve remarkable advances in a very short space of time in so many different areas. For example, clients have been able to:

- Focus their time on what will make the biggest difference to their success.
- Reduce the risks of failure and be more prepared if those risks start to materialise.

- Dramatically increase buy-in, getting more people to support their work actively.
- Deliver their projects in record time, with fewer problems, and enhanced benefit realisation.
- Pull out of the detail so they could take a more strategic view.
- Achieve much more with much less.
- Become more confident, assured and less stressed.
- Attract the attention of the talent spotters at the top of their organisation.
- Move forward their career, sometimes several steps at a time.

The Stakeholder Influence Process is also useful in a very wide range of pursuits. Over the last ten years, I have seen it being successfully used by clients who are:

- Delivering programmes and projects (big and small).
- Launching new products.
- Starting businesses.
- Managing crises.
- Implementing change initiatives.
- Turning around failing businesses.
- Accelerating their careers.
- Managing suppliers.
- Introducing new systems.

Basically, anything you need to achieve which requires determined effort over a period of time and involves lots of people, is likely to be aided significantly by the Stakeholder Influence Process. Recently, I saw one delegate on a workshop use it during the coffee break to develop a strategy for improving the relationship she had with her mother-in-law!

Not Applicable Here

If you've got it, and are convinced that the Stakeholder Influence Process could work for you in your situation, please feel free to skip over this section — it's for those who don't think it could work for them. These people generally fall into two main camps:

1. External contractors who have been engaged by an organisation to deliver a project.

2. Internal project managers who are responsible for small projects, or even, those who are not formally project managers.

Below you will find some challenges to your mindset, approach and reticence, together with some ideas and suggestions.

External Contractors

"This won't work for me because I have been engaged by an organisation to deliver on a project. They have given me clear terms of reference and stepping outside of that is not what I am being paid to do. They are responsible for creating the conditions under which I can deliver on my contract."

You are absolutely right if you think this. What it means is that you have taken on an assignment without negotiating the terms that would have given you the power, access and/or permission to make it happen. You have relied upon their promise that they will remove all the obstacles, assign all the resources and that this is their mission critical project. And I don't blame you at all, they probably believed they would be able to do all of that too.

The fact that everyone behaved with the best of intentions doesn't alleviate the pain of the problems that often follow poorly constructed and negotiated terms of reference. Who was responsible for it will always be a matter for debate; however, who takes responsibility for putting things right is a choice — and that could be your choice.

The Stakeholder Influence Process will work perfectly in your situation if you want to make a difference and can get yourself into a position where you can apply it. In order to do that you have to make a decision to do it, and then decide how to focus your influence. The next chapter is devoted to helping you to define the focus for your influence. In your position, it might be that you need to:

o Influence your client to redefine your role. Instead of simply being the project manager, you take on responsibility for engaging with internal stakeholders. Or perhaps, to take a more active role in the steering committees. Taking it one step further, maybe what you really need to do is become a consultant project manager.

o Influence your client to deploy the Stakeholder Influence Process, especially the parts about understanding the political agendas and the bigger picture. Often, those responsible for external contractors are nowhere near savvy enough to understand and handle the internal politics.

- Influence your client to take action to resolve specific problems that are holding back your delivery. Indeed, you may focus on influencing them to work with you on the resolution.

Of course, if you are in a position where you are just about to agree terms, stop and think about it now. Yes, I know you are keen to get the assignment, but at what cost? There is little point taking on something that is programmed to fail because the powers are not vested in the right hands.

Many external (and internal for that matter) project managers sit behind the excuse of their terms of reference or brief. The reason why they are not delivering is because they do not have the permission to fix the problems. For my money, I would sooner work with a project manager who has the insight and skill to alert me to the problems that I am beset with, and help me to resolve them for the success of the project I am responsible for. Those are the project managers that come to the top of the list for re-hire and at the best rates!

Incidentally, the reference to consultant project managers above was deliberate. In my experience, external consultants behave very differently with their clients than contract project managers. This is another example of adopting a different perspective to your role. Consultants usually work with their clients in the political environment to achieve a mutually desirable outcome. Remember the relationship perspective from the last chapter?

The bottom line is if you know what is wrong, and your client doesn't, or won't acknowledge it you have two choices. Either you do something about it, or you don't. The Stakeholder Influence Process can be applied if you decide you do want to do something about it.

Internal Project Managers

Regardless of the size of your project, or even if it is too small to be called a project, the Stakeholder Influence Process can help you to make things happen. At a more junior level, some believe that it won't work for them because they do not have the access, skills or confidence to implement it. If you're thinking that, you're probably right. However, you don't have to settle for this, you can change it.

In the next chapter, I will be challenging you to think about what you need to influence in order to break through some of the problems you are experiencing, and to help you to become more successful. As you read on, bear in mind that:

- Whatever your current position, you have the right and responsibility to attempt to improve things. That is your obligation to your employer, to yourself and your family.

- Everyone has to start somewhere. Applying the ideas in this book will become easier with practice.

- Start small. You don't have to apply the Stakeholder Influence Process to massive projects or challenges right away. Begin modestly, become familiar with the process and mindset described later. Then expand your ambition as your confidence grows.

- Relationships are key. If you are not yet ready to push hard to influence things, focus instead on improving the quality of your relationships. As trust grows, your ability to influence them will also grow — along with your confidence.

Here's a final thought for you. If you are at an early stage in your career as a project manager, take heart from the fact that it will be far easier for you to adopt the mindset necessary to make these processes work than it is for those with many years of experience. They have learned how to be successful in the old world and will struggle to look at things differently. Old dogs, new tricks? To be honest, and they won't like me for saying this, you have a golden opportunity to break through and overtake their track-records. Don't let them keep you down. And if you're an old dog reading this, don't say I didn't warn you.

Remaining Realistic

While all of the above sounds great, it needs to be recognised that the Stakeholder Influence Process is not going to solve all of your problems immediately. What it will do is help you to identify the actions you need to take in order to start making progress on your challenges. You still have to take responsibility and implement the actions. You still need to exercise skill, persistence and creativity in arriving at the solution. Without the Stakeholder Influence Process, people often struggle to make sense of what is going on. They get stuck in the detail, buffeted by the problems and issues, and are unable to make sense of what is happening to them (or their project). This process is the beginning of the end for these problems.

Another factor that needs to be stressed is that for some readers the benefit will be unexpected and, at least in the short term, a little uncomfortable and stressful. Occasionally, when I have coached people with this process, they have realised that the goal they are striving for is simply not going to happen. Maybe the powerful forces moving against them are simply too much, the opposition too strong, or the project they are working on was ill-conceived and really should not be implemented. Going through the process can become a little emotional, but at the end of the day there is little point working on something that is doomed. Eventually, the process gave them the confidence and the thinking, to engage their stakeholders in healthy and objective debate and bring forward the decision to cancel their project. Without doubt, these decisions were tough. However, what these individuals managed to avoid was many hours of agony and wasted time, effort and money.

Those who had the courage of their convictions benefited greatly in the long run because their organisations realised the value that they brought. Although their projects may have been cancelled, they quickly became eagerly sought after for other critical initiatives. In many ways, what they had discovered is the power that often comes from showing such commitment to the organisation that they were even willing to stake their own future on finding the right solution for the organisation. A rare quality indeed!

However, these sorts of unexpected benefits are very much in the minority. Most people can protect their projects from sabotage, conflicting priorities and agendas. Basically, they move their work forward with much more support and safety than their competition — simply by applying the Stakeholder Influence Process.

There are no magic pills, panaceas or quick fixes here. But what this approach does do is help you to quickly find the action which needs to be taken, the arguments or debates which need to be conducted, so that you can move forward and overcome any opposition which may be waiting to thwart your endeavours.

Exploring Your Motivations

The bottom line is that in this book I am going to be asking you to invest time, effort and energy. How much depends on how close you are already to the approaches that I am going to advocate. Regardless, it is likely that you will experience a little strain in some way, and the only way to move through that, and to be able to persevere with the implementation, is to maximise your motivation to do it.

This exercise is designed to stimulate your motivation and prepare you to give this book the time that it, and you, deserve.

If you can easily handle the challenges you identified earlier:

- How will this improve your performance and results?
- What difference could it make to your state of mind?
- Will there be health benefits? If so, what?
- What knock-on impact will there be to your personal life?
- How might this improve your career prospects?
- How will your organisation benefit?
- What will it do for your reputation?

Motivation is a tricky part of psychology. At all times, it is working in the background unseen. For you, engaging in this book, the problem this presents is that below the surface there are a lot of other motivations competing for prominence, such as going home at a reasonable time, relaxing and chilling out, or responding to an irate stakeholder.

Generally, the motivation that wins is the one which shouts loudest, unless you do something about it.

If you want to realise the benefits of becoming proficient at handling the challenges you are facing as a project manager, then you have to make this book a priority. That is not to say other things are not important too, just that you need to make it more conscious.

Here are a few suggestions to make achieving these benefits more likely:

1. Don't skimp on this exercise. Take your time and write down as many benefits as you can in your notebook.

2. Make sure and keep this list close to hand so that you can remind yourself of why this work is important.

3. Don't consider this to be beneath your long years of experience. This stuff works and, if it is important that you gain the benefit, just do it.

4. Add to your list each time you review it.

Key Points

- Many of the challenges faced by project managers can be significantly alleviated or removed altogether if the project manager gains sufficient influence.

- The Stakeholder Influence Process is a simple way to stay focused on the important questions you need to be asking and the actions you need to be taking, in order to achieve the influence you wish to achieve.

- This book is about taking practical action and learning as you go. Although you are strongly advised to create the time to study it in detail.

- Adopting a more influential approach is a choice, your choice. The alternative is to continue getting what you've always got.

Suggested Actions

- Make a commitment to becoming more influential.

COFFEE BREAK

Taking Control

The purpose of these breaks is to give you a chance to relax and ponder some of the common concerns of project managers. They are not necessarily related to the content you have just read, or are about to read. Try to treat each one as a positive encouragement to take some action.

When things are not going the way you want them to go, to what extent are you responsible for this? It is easy to make excuses and place responsibility, especially for problems, with someone else. This helps because:

○ It allows you to focus on the things you can do something about.

○ You can avoid feeling like you should be doing more about the problem.

○ Your team will not blame you for the challenges they are facing.

However, is this what influential project managers should be doing? I regularly notice a lack of readiness in project managers to take on responsibility for making things happen. Deciding to act on a problem or challenge doesn't mean that you are the cause. I would argue that the continuance of the problem or challenge is down to you. Influential people seek out what needs to change, and then take action to bring about the change.

But, never mind what I think, what do you think? Are you making excuses for the problems you are experiencing? Could you be doing more to bring about a resolution? Are you taking enough control of what is happening to you and your team?

CHAPTER 3

Step 1: Focus

As a project manager, you are familiar and practiced at focusing on goals and deliverables. However, in order to achieve those goals, in a political environment, what have you got to influence?

This chapter will help you to:

o Review and reflect on the challenges, issues and problems which you would like to work on.

o Decide the most important thing to focus the Stakeholder Influence Process on.

o Increase your clarity on exactly what needs to change in order to achieve your goal.

o Develop a list of other goals that you can work on as time allows.

o Raise your motivation to make things happen.

o Focus more on your work as an influencer.

If you're in a hurry:

- Spend a little time reflecting on *Establishing Your Priorities*.

- Read the section on *Being More Specific*. Make sure you can document evidence criteria for what you wish to achieve.

- Make sure you understand the key points at the end of the chapter.

Unless you know exactly what you are shooting for, you will lose ground, miss opportunities and struggle to get buy-in. If you are clear, you'll move much faster, save time and get even better results. You need to get focused not only on your end goal, but also on what you need to influence to get there. The first step in the Stakeholder Influence Process is to clarify what you want to focus on.

Much of what follows will depend on your current position, challenges and ambition. At a more senior/experienced level, you may have a fairly clear notion of your longer term life or career purpose. When you read the pages that follow, use it to reflect on where your priorities lay in helping to further that cause.

Alternatively, you may be beset with difficulties and challenges right now. These often cloud the longer term focus and need to be dealt with. By bringing these under control and making progress on them, you will, over time, create space for extending your ambition and looking at the bigger goals. In this situation, use the pages that follow to get moving on what needs to be resolved right now.

Exercise: Establishing Your Priorities

Pause to consider all of the important things that are going on in your life/work at the moment. Think about:

- What targets and projects have you got at work this year?
- What are your personal goals for this year? Next year?
- Where are you hoping to take your career?
- Are you facing particular difficulties or issues at work (or home) at present that you'd like to resolve?
- What do you want to achieve in the next couple of years?
- What goals are there beyond the ones right in front of you?

This is a warm-up exercise to get your thoughts moving and considering your bigger picture. In a moment, you can start to work on developing the focus. This action can be very useful in making sure you've not missed anything critical. If you've already done this general thinking, no problem, just move on to the next section. Otherwise, put the book down and muse a short while.

Deciding on Your Focus

Once you've taken a little time to reflect on this, you need to settle on one or two topics that you can use this book to move forward on. Ideally, you're looking for a goal that:

- Is of critical importance to you.
- Can motivate you — or even get you really excited!

- Will stretch your skills and involve a fair degree of influencing to make it happen.
- Ranks high in your list of priorities overall when deciding where to spend your time and energy.

Bear in mind that the goal you want to move forward could be a shared goal — it doesn't have to be something that is only important to you. In fact, if it is a shared goal it will be even more suitable for the Stakeholder Influence Process. You can all explore, learn, and play your part in realising the objective. You'll also be helping your colleagues to learn how to use the process that will be beneficial in their wider work too.

You may also decide to work on somebody else's goal. By that, I mean that you may not have formal responsibility for it, but if you feel strongly that something needs to happen, you could choose to make it your business to push it forward. But before you rush off poking your nose into other people's business, stop to consider how your action may be perceived — not only by the person who should be pushing forward on that goal, but also by the stakeholders who are connected with it. Some may wonder about your motives, so think about doing some careful positioning before you go too far, too fast. At the end of the day acts like these are sometimes necessary.

Goals that tend to work best in this process are the ones that:

- Are a top priority for you in your work.
- Require many people to think, feel or act differently.
- Need to overcome tough opposition from others.
- Require a plan of action over several months.

- Are important to other people as well.

Put another way, a goal that is rarely suitable is one that:

- Only needs one person to say yes.
- Could be achieved in one meeting.
- Really belongs to someone else.
- Has an easily understandable logic which everyone can agree to.
- Is more of a wish than a key part of your life or work.

Of course, if the one person you need to say "yes" is inaccessible to you because they are too high in the organisation, this could be very applicable to this process. In this case you will need to find lots of others (stakeholders) who could influence on your behalf or open up the opportunity for you to speak to your target.

To decide on your focus for this process, start by considering your list of potential goals and choose one that you want to make some serious progress on. For instance, which goal is:

- The most exciting?
- Causing the most stress?
- Likely to help you to take the biggest step forward?
- Going to make lots of other goals easier?
- The most motivational?

Of course, you can use the Stakeholder Influence Process for more than one of your goals, but make sure to consider them separately to avoid your influencing effort becoming confused. You will definitely need to use a different stakeholder map when you get to the Analyse step.

The activity so far in this chapter should have helped you to review and focus what you want to make progress on. Yet for this process to yield the maximum return, you need to think in terms of what you have to influence in order to achieve your goal. For this reason, I use the term Influencing Goal. Sometimes it is more effective if you focus on the influence you need to achieve than the hard tangible target you are aiming for. Achieve your influencing goals and you'll hit your main goal.

Turning Goals into Influencing Goals

Most people are familiar with the need to define and describe their goals. Usually this is a part of the line management process. Specifying unambiguous targets, deliverables and achievements is the life blood of performance management. These are what I usually refer to as hard goals, and quite likely they are the goals you came up with in the previous section. For example:

- Complete Project X by the end of September 20xx.
- Secure £250,000 additional funding for your project.
- Increase Market Share to 12% next year.
- Achieve a clear go/no go decision at the next Steering Committee meeting in July.

What these hard goals usually miss is the need to bear down on the influencing work that has to take place in order to achieve them. Hence, I like to stretch my clients to begin specifying the often subjective nature of the influencing work they have to do. These are the influencing goals.

For each hard goal they are working on it is likely that many different influencing goals will make a contribution. For each one, the Stakeholder Influence Process can be used fully. There are times when the distinction is not necessary, although in most cases it is really important.

In order to begin turning your goals into influencing goals, use the following definition of influence:

> Influence = People acting, thinking or feeling differently.

There are a couple of interesting implications of using this definition.

Firstly, notice that it is neutral of intent or means. For influence to have happened, people need to act, think or feel differently. They either did or they didn't. It does not define the conditions under which they were influenced. Some may have been forced; others may have been chomping at the bit to act differently.

If you wish to add an ethical dimension to this, ideally you would be striving for people to *want* to act, think or feel differently. They have considered the desired change, have considered their options and realise that it is in their very best interests to make the change.

However, that is not always expedient or even possible. If you are not able to get people wanting to do what you want them to do, aim next for people being *willing* to do it. This means that they are prepared to do what you want them to do, perhaps without argument or debate. The emotional element to being willing can, of course, vary greatly. Some may grudgingly say *yes* because they cannot find any other reason to justify opposition. Alternatively, they may have assessed the situation and while they are not really wanting to do it they have decided that on balance, their best interests are served by doing what you want. Their agreement may be due to negotiation, compromise, strategic calculation, or fear.

Be careful that you are not forcing people into *willingness*. This could be catastrophic for your longer term goal. Turn your back, and they will revert. Or, they may find more powerful friends to back up their opposition to what you are trying to do. Neither of these is a good outcome for long term success, if that is what you are aiming for.

In my experience, there is far too much influence created on the basis of people being merely *willing* to do what someone else wants them to do, rather than genuinely buying into the change. That presents a significant risk to implementation of any action, project or idea. Thus, while the definition will remain neutral, I would heartily encourage you to work hard to get people to really want to do what you want them to do, and if not, at least be transparent in their willingness to do what you want.

Another factor that needs to be borne in mind is that influence goes beyond mere action. Often you need to move hearts and minds and, therefore, might need to change the way people feel or think. A common example of this is getting them to feel positive about a forthcoming change in procedures or systems. Alternatively, you may need to influence a group of sales people to remember to make a decision about involving the marketing team at the right stage of a particular process. You might not need them to involve marketing on every occasion, just make sure that they think about it and make a clear decision about the need to involve marketing or not in any given situation.

You will also notice that my definition includes a change of some sort. Strictly speaking, this is not always the case. Sometimes, it is necessary to ensure that your stakeholders continue acting/thinking or feeling in a particular way. However, most of the time you will need to work on creating a change of some sort.

The final point is that this definition refers to "people." The reason for this is that the Stakeholder Influence Process will generate maximum benefit when it is focussed on helping you to build strategies to move the masses, rather than just individuals. It can be useful with individuals, especially if they are very senior people. If so, you will need to create a campaign of action to get them to endorse your goal with the help of the influence process outlined in this book. That said, most of the time you will be working to influence many people as part of your goal.

With this in mind, the next question to answer is:

> What do you want/need to influence
> to dramatically improve progress
> towards your chosen goal?

When you can answer this, you can begin to define influencing goals. With your main goal in mind, consider:

- What could you change which will have a big positive impact on your success? Could, not can — this is not the stage to apply your talent for realism.

- What are the major obstacles standing in your way?

- Do others hold particular attitudes that stand in your way or are working against you?

- What behaviours need to change to move forward quickly?
- Who do the decision makers turn to for advice?

Don't neglect to consider that it may be something outside of your direct control. Many times it is necessary to influence a number of different parties to work together, reach an agreement, or stop fighting with each other. In this case, you can still define an influencing goal and attempt to make it happen. If something is standing in your way, it is something you should decide to take action on.

In effect, what is happening here is that for each main goal you are defining a number of important influencing goals. In your work as a whole, you could visualise the structure thus:

>Main Goal 1:
>>Influencing Goal 1.1
>>
>>Influencing Goal 1.2
>>
>>Influencing Goal 1.3
>
>Main Goal 2:
>>Influencing Goal 2.1
>>
>>Influencing Goal 2.2
>
>Main Goal n:

For example, if your main goal is to secure additional funding for your project, you might identify influencing goals such as:

- Get client/customers bought in to the additional benefit they will gain from extra funding.

- Senior management advocate that this project should be made a top priority.
- Remove Finance opposition to increasing project funding.
- Steering committee agree the additional funding should happen.

Similarly, if your main goal is to "achieve sales revenue of $650k," an influencing goal which is likely to make a significant contribution to that achievement could be to "get the Marketing Board to sign off your sales strategy." This focuses on people doing something different — signing off your strategy. It is quite likely that each significant goal will have a number of main influencing goals moving you closer.

On really big goals that are running over a number of years, it is entirely possible that you may need to further divide each influencing goal into sub-goals. Before you get too carried away, make sure and check that each influencing goal you are contemplating is capable of standing up against the criteria summarised in the previous section.

The skill here is in being clear about what you want to achieve, getting very specific on exactly what you need to influence in order to achieve it, while also remaining pragmatic that it will be able to get you moving fast. It is impossible to focus on everything you need to influence, so make sure and identify the things that will have the biggest impact.

When you are applying the Stakeholder Influence Process for the first time on a given goal, it might not be possible for you to stretch goals into influencing goals because you do not yet know enough to be sure what needs to be influenced. In this case, just proceed with your main goal as the focus for the process and cycle back later as more becomes known. Before you do though, at least take some time to attempt to define them as this will help to you to pick up speed later.

Goals have a habit of evolving as you work on them. Sometimes you aim too big, sometimes too small. Then again, you may realise you are heading in the wrong direction, or perhaps, shouldn't even be trying. Make a habit of using this process to keep you asking the right questions along the way (I'll talk more about this is the Maintain step).

Becoming More Specific

A final piece of thinking in this step of the Stakeholder Influence Process is to consider how you will know that you have achieved the desired influence. Not all influencing goals are clear and unambiguous. For instance, the one about influencing senior management to advocate the project as a top priority may be very difficult to evidence conclusively.

I'm not going to patronise you by talking about SMART goals and things like that, but it is important that in your own mind (or even better, on paper) you are able to be crystal clear about what success looks like. The clearer you can be, the better. Unless you get specific, you may simply have some aims. Nothing wrong with that but if you want to move faster, get clearer.

By their very nature, influencing goals are often soft or vague. In these cases, it can help you to define a set of evidence criteria you can use to assess progress and success. When you do this, identifying the actions to achieve them also become very much easier (but more on that later in the book).

For example, if your influencing goal is to gain senior management advocacy for your project, evidence criteria may include:

- The CEO reports that Project X is vital to delivering the organisation's strategy.
- All directors insist that none of their projects are hindering Project X.
- Project X is a specific agenda item on all Executive meetings.
- New Project X team members are all rated as "top talent" by the performance management process.
- Other projects are regularly passed over to give air time to Project X.

You will be right in thinking that none of these are conclusive evidence in their own right; however, in combination they make a compelling case that you have achieved your goal. You probably don't have time to spend being perfectionist on this one, so define a good list of the evidence that you would expect to see if things are moving in the right direction.

Another reason why this is a good idea is that on occasion you may achieve your influencing goal in an unsafe manner. For instance, imagine that the CEO tells you that the senior management are all advocating and supporting your project. Now, I don't want to cast aspersions on your CEO's judgment, but in many cases these types of assertions are far from reality. Often, this is because the opposition is hidden from the CEO, or the CEO may be compelling or forcing agreement. When the implementation begins, the real picture will emerge soon enough. So, defining a clear set of evidence will help you to verify the real position, and if it is not coming through, you can make decisions about what to do in order to make it happen.

Over the years, I have found the Focus step in the Stakeholder Influence Process to be one of the most important. When clients are not moving forward quickly, it is usually because insufficient focus has been developed.

The time and effort that you devote to clarifying your focus will be well rewarded. I remember one director I was coaching who paused to consider what he needed to influence in order to achieve his goal. He soon became clear and went off to engage a stakeholder who was against his idea. Later, he reported back that the stakeholder had easily agreed when he was given a clear request. Apparently, he had been fantasising about lots of things my client wanted which were actually untrue. Assumptions have a lot to answer for.

Before moving on to the next step, it is worth returning to a topic covered in the previous chapter — motivation.

The Psychological Path of Goals

As you work on your goals, your mind is likely to transform your beliefs about your goal. Being aware of the stages your beliefs move through will help you to find ways to accelerate your progress. Although it is possible to skip stages, it is rarely advisable. Missed stages create problems later.

Here are the stages:

1. **Want.** The first inkling of an idea, the first rousing of desire that you want to change something. Many get stuck here, thrashing around with often conflicting notions of what they want to

make happen. To move to the next stage, get specific: exactly what do you want to achieve and why? What evidence will there be that you have achieved your goal?

2. **Must.** During this stage, your *why* gets firmer; motivation and commitment rise. The key to getting through this stage is thinking of as many consequences as you can. What will you gain from success? What will you lose if you fail?

3. **Can.** This is where you build belief that it is possible to achieve your goal. This stage is far easier to pass through if you have a strong must. If you don't, you'll probably just find all the reasons why you can't achieve your goal and then give up.

4. **Will.** As with the other stages, this builds naturally from the strength you gave to the previous ones. You know you've arrived here when you suddenly realise your can has become an inevitability you have made the decision to make it happen, and you are totally confident of success.

5. **Did.** Right, now on with the next goal with greater confidence in your ability to make things happen.

Plan well for these stages and find ways to move quickly and confidently through each stage. The trick to this is making sure that you invest sufficient time and energy into each stage rather than skipping them. Don't miss one out just because you think you can. Each stage thoroughly completed adds greater insurance that you will complete the path.

Incidentally, these stages are not completed when you have simply thought them through logically. They need to progress at an emotional level as well. It is easy to work through all the benefits and rationalise that it's a good thing to do, but only when your heart moves and your gut tells you it is going to happen will the stage be completed fully.

You also need to recognise that others have to go through these stages relating to your goal too, especially those who are closely connected to it like team members, key stakeholders and customers. Again, plan action to move them through the stages healthily. In many ways, this is an alternative way to look at stakeholder engagement and buy-in.

Key Points

- You need to use the Stakeholder Influence Process to focus on the things that are most important to you.

- Use it to resolve the challenges you are facing today first. Once you're making good progress on these, then extend your time horizon.

- Your focus is your choice.

- Influence = People acting, thinking or feeling differently.

- By focusing on what you need to influence, you will easily identify simple actions that you can take to influence your stakeholders.

- Remain alert to your progress through the psychological process from *want* to *did*.

Suggested Actions

- Keep a record of your goal ideas. If you are not focusing on them today, review often as you get more things done.

- Consult with key stakeholders about what is important. They may have other ideas that you have not considered.

- Do the exercise to define your goals, influencing goals and evidence criteria. It works.

- Make a commitment that you will always aim to get people *wanting* to do what you want them to do.

COFFEE BREAK

Lack of Authority

Once upon a time, project managers were assigned to projects and were given a dedicated team to make things happen. People were seconded to the project from different parts of the organisation and other people backfilled. The project manager was able to plan with the certainty that control of these people was theirs.

Not anymore. In the vast majority of cases, projects are staffed by part-time resources. People are allocated for a percentage of their time while their main bosses expect them to continue to do exactly the same job for them. That they have been assigned to another project for 50% of their time does not seem to lessen the need to do their existing job.

o   How are part-time resources impacting your project?

o   What can you be doing to strengthen your relationship with them?

o   Are there benefits to you because they are part-time?

This is the reality for most project managers in big organisations. It means that in order to succeed, you will need to be brilliant at managing the relationship with your people. It is vital to build high levels of trust so that they level with you about what they are really able to do. At least then you will know what is going on and may be able to negotiate. Understanding them as an individual is critical in order to find the keys to motivation. Everyone is different, and project managers need to know how to unlock the discretionary effort of their people like never before.

CHAPTER 4

Exploring Organisational Power and Influence

Since the political perspective is based on the notion that all, or most, projects are designed to adjust the distribution of power within an organisation, it is vital that you understand exactly how it works. This will take it from a mysterious political world into the realm of objectivity that you are probably more comfortable with.

This chapter will help you to:

○ Learn what influences people and makes some people more influential than others.

○ Understand the mechanics of power and influence.

○ Take stock of your personal power and how this is affecting your influence.

○ Prepare you for developing greater power and influence.

If you're in a hurry:

- Skim read the *Sources of Power* section. Most of this you probably know and all you need is a quick reminder.

- Read the *Principles of Power* carefully. This illustrates the mechanics of influence and will help you to apply this to the world around you.

- I know you're in a hurry, but make sure and carefully study the section on *Group/Organisational Power*. This lies at the heart of many of the challenges you face, and until you understand this, your progress will be limited.

When individuals make decisions about what they should do or how they should change, one of the (often unconscious) influencers is the presence of powerful people who could help or hinder them. Everyone has an agenda, and it is natural to think about how others may lend a hand, or get in the way. Understanding how power affects these decisions will help you to explore simple ways to create the influence you need.

It is not a simple case of understanding how individuals are influenced; you also need to grasp how things work when individuals come together in groups or organisations. Groups are complex places at first sight, and this increases with size, geographic spread, and business diversity. The arrival of matrix structures and cross-functional working has made organisations more integrated and nimble, but at the same time has added another layer of complexity.

Once you can work out why people do some things and not others, you will start to discover easier routes to getting things done. Who to involve, who you can ignore. Which buttons to press and levers to pull. Trouble is, there is no rule book as these things tend to emerge within the culture of the organisation. They are not easy to see. Of course, there will be a sign-off process, but usually these are just rubber stamping exercises — the decision has already been made within the social fabric of the organisation.

The quickest way to learn how all of this is working, is to study the subject of power. When you've got a clear insight into how power works you will be able to begin to decide what you need to do in order to become more influential.

Sources of Power

Power is not complicated — it is actually quite simple. The concept is regularly spoken of, but rarely understood at a practical level. Jeffrey Pfeffer (author of *Managing with Power*) usually defines it as:

> Power = The capability to create influence.

In the last chapter, I used the following definition of influence:

> Influence = People acting, thinking
> or feeling differently.

Consequently, to understand how to acquire the capability to influence, you first need to know what influences or motivates people to change. A great deal has been written on this topic; much of it has been over-complicated by well-meaning authors. My preference is to keep things simple and accessible for those who need to move fast, so I have organised the things that influence into seven broad categories or sources.

These sources are not intended to be mutually exclusive, so do not waste your time getting too obsessive about what goes where. Just go with the flow. Once you've read through the different sources, complete the exercise to make them relevant to your world.

Credibility

The power derived from your professional standing and expertise.

Credibility is the reliance awarded to someone based on trust and perceived expertise on the topic being considered. When making decisions, most people get input from others, and they have to decide how much reliance to place on that input. Information from highly credible sources is more likely to be accepted at face value and not challenged, questioned or verified with another source. If someone doesn't think you are credible, they probably won't believe what you have to say.

For example, things that may render someone credible include:

>Qualifications, Awards, Endorsements, Experience, Knowledge, Track-Record, Results

Character

The underlying traits, values and beliefs which shape your behaviour.

It is difficult to estimate how influential strength of character is, but it is big. It complements credibility well, and if you have generous amounts of both, you're well on your way to becoming very powerful and influential.

In effect, this is concerned with your internal world and how it manifests in observable behaviour. The old phrase *character building* seems to have been forgotten in leadership development, but many years ago it was top of the agenda for all those concerned with developing leaders.

Character traits that often boost influence include:

> Tenacity, Fortitude, Integrity, Engaging, Tough, Energetic.

Presence

The impact you create and the feelings you stimulate when people meet you.

Much is made these days of Executive Presence, and with good cause — because it is a source of considerable power and influence. It is usually described vaguely as a collection of characteristics that combine to create an aura around the executive, making them immediately noticeable. When they glide into the room, everyone immediately feels and responds to their presence.

Presence is created by the external manifestation of the internal state of high levels of confidence. Therefore, to have presence, the individual needs to really believe that they are in the right place, with the right skills and capabilities to do what is expected of them, with ease.

Consequently, people with presence are usually:

> Poised, Calm, Fluid, Good-natured, Humorous, Charming, Confident

Position

The roles you play and how you manoeuvre yourself into the limelight.

To become powerful, it certainly helps to get yourself into a good position. Position means much more than simply being given a particular role or title by your organisation, even though this type of position is the easiest to recognise.

Beyond this, you have people who are given temporary positional power as project leaders, managers or sponsors. All of these have been given responsibilities by the organisation and the power to do certain things. Other powerful positions sit to the side of organisation control. Union conveners, spokespeople and staff representatives can all exert significant influence on organisational life.

Then there are the myriad positions within the social fabric of the organisation. Party organisers, newsletter writers and "go-to" people all have positional power. This can even go as far as the company joker, the life and soul of the Christmas party who is in a position (attention) where they can flatter or ridicule with equal mirth.

Position is usually reliant on control of one or more of the following:

Money, People, Knowledge, Information, Information Flows, Attention.

Connections

The network of relationships you have around you and your work.

Without doubt, connections can make you more powerful. By connections, I mean the network of friends, associates and contacts. All of these provide you with the capacity to influence people to support the achievement of your goals. In turn, this enhances your credibility because of your strong track-record and results. Thus, a virtuous circle is created.

On top of this, having a strong network of connections can also help increase your influence in a number of other ways. It can make it more likely that others will favour you because you have the potential to introduce them to new people, or the opportunity to put in a good word for them (or a bad one).

However, it is not just a matter of collecting names in your address book. Quality counts, and increasingly so.

Skills

Those exceptional abilities you have which enable you to get things done.

People don't get ahead and become highly influential by being the same as everyone else. This source points to the particular skills which you have that really stand out. You are not just good at these; you are exceptionally good, and others know it. These skills need to be relevant to your role and the people around you.

To be influential and a source of power, they need to be really good. So good that they are capable of influencing people without you even using them. For instance, someone with superior negotiating skills is likely to be influencing others simply because they are due to attend the meeting.

Other skills that have the capacity to influence without being used include:

Approachability, Problem-solving, Critical-thinking, Persuasion, Repartee.

Agenda

The issues and priorities you focus your attention on.

What is noticeable about all powerful and influential people is that they have very clear opinions about a whole variety of things in the wider organisation. They have also chosen to focus their agenda in a particular way. Their focus may be orientated towards fixing a certain problem or issue which the business faces. Or perhaps they have decided that the wider population need to view things in a specific way. Whatever it is, they have clarity and most people know what this is.

The ability of the chosen agenda to influence will vary depending on the strategy of the group or organisation. If you know what these are, it is relatively simple to jump on the powerful agendas and gain more influence.

Powerful agendas running in many organisations include:

> Cost-cutting, Market growth, Centralisation, Globalisation, De-layering, Standardisation.

These seven sources of power combine to create the reputation of the individual concerned. Your reputation creates expectations in the minds of others about how you behave, what you do, and how you may be able to help or hinder others in the pursuit of their own agenda. I'll talk more about this shortly when I share the underlying principles, and again in the final chapter.

What Power Means to You

Pause for a moment and do the exercises below. It is easy to read and agree with what I have written above; however, if you want to get the most from this, you need to spend some time developing your awareness of how power is influencing those around you, and also, what makes you powerful. The purpose of these exercises is to help you to make this topic more relevant to you and start you on the path of becoming more powerful.

Spend some time reflecting on your answers to the following questions:

- What does credibility mean to you? What would convince you that someone is credible? How would you quantify credibility?

- How does this change for people in different roles? What makes a credible boss? Or a credible accountant?

- Who are the strong characters around your workplace? What is it about their character that you admire? What are you less enamoured with?

- Think of someone you know who has presence. What is it about them that makes you think that? Why are they able to create the impact that they do? Does everyone think that way about them?

- What positions are you aware of around your work that are capable of influencing many people? How do the positions vary in terms of formality? How do the different positions compare in their ability to influence?

- Who are the best-connected people that you know? How does this help them to influence? Is there anything special about the connections they have?

- What skills is your organisation desperately short of? Or, what skills do people in your workplace really take notice of? Who stands out because of their exceptional skills? What are they, and why do they make these people distinctive?

- And finally, agendas. What are the powerful people you know well always talking about? What is it that they care most about? Why are they so passionate about these things? How does this fit with the strategy of your organisation? How does it fit with your agenda?

I make no excuse for the amount of questions above — they are important. In time, they will become part of your everyday thinking.

It would be a good idea to use your notebook as you ponder these questions:

- Which sources of power do you tend to make most use of when you are influencing others?
- For each, write down exactly what you mean by it. What is it about say, your credibility, which is working really well for you?
- Think of someone who influences you. Which sources are working for them when it comes to influencing you? What is it exactly about those sources that can move you?

The Principles of Power

If you want to become more influential, it really helps if you know why the sources of power are creating influence. Once you've understood these principles, you will be able to economise on your work as an influencer, and gain a great deal more influence.

To make this nice and easy, let me give you a simple illustration. Imagine for a moment that you have just won the lottery and now have a bank account full of money. This gives you a tremendous capacity to influence. Immediately your bank manager is going to be far more attentive, hoping that she can encourage you into making some shrewd investments. Similarly, you'll probably begin to discover lots of best friends you never knew you had. Many people will be acting differently around you, having been influenced by the presence of all that cash.

In this example, you have positional power, being the lucky owner of a bank account full of money.

Now, here are the five general principles which apply to how power is creating influence.

Consequences

Power creates waves and needs to be handled with care.

In the lottery example, the consequences are going to be immediate. Not only will other people be noticing it and perhaps viewing you in a different light, you will also be transforming as suddenly you have no more financial worries — whatever will you spend it on? This doesn't just happen with money, all forms of power can quickly go to your head, warping your behaviour and corrupting your integrity.

From the other direction, if you suddenly get a promotion, other people may start queueing up at your door rather than making you queue at theirs.

Calculations

Power is all about individuals making personal decisions in order to satisfy their needs and wants.

Remember your friends, well they will certainly be thinking about you and your new found wealth. They will also be thinking about their own worries (agenda) and may start to hatch a plan to alleviate you of some of that money. Better friends will be calculating how to save their friendship with you as money often spoils things like that.

Whatever the source of power noticed, people will be calculating how it will and could affect the progress towards their own goals.

Supply and Demand

You will be powerful if people want what you have, and especially powerful if they can't get it elsewhere.

Power works according to the laws of economics. Those friends of yours who are in most need of financial assistance will be the easiest to get dancing to your tune once you've told them about all your money. This will be especially so if they have just been turned down by their bank for a loan. They are in demand, and you have the supply.

All forms of power work this way. Those who have control of the things that people need see their influence soar. To capitalise on this principle, all you have to do is make sure that you have control of what everyone wants and cannot get anywhere else.

Perception and Reality

It is not so much about what power you have, but what power people think you have.

Their perception of your power may vary from reality. They may know that you've suddenly come into a lot of money; however, they may not know that you need to use it to pay off your mountain of debt, or have already committed it elsewhere. Similarly, if you had managed to keep everything quiet, they would have the perception that you are still poor and would continue to ignore you.

Whenever there is a large gap between your actual power and the power people perceive you to have, there are likely to be problems just around the corner. If you are less powerful than people think, at some point this is going to become obvious, and you will lose credibility by being unable to meet their expectations. In the opposite case where you are more powerful than they think, you'll either not get as much support from them as you could, or they will be in for a nasty shock when they do realise.

Make sure to manage these gaps carefully.

Utilisation

Your power and influence will grow if you use your power.

The final principle is very important in the long term. If you don't use your power, you'll lose it, or at least, its capacity to influence will dwindle. How long before your new friends realise that you may have lots of money, but there is no way you are going to share it with them?

Aside from communicating to others that you have power and that you know how to use it, exercising will increase it too. In a similar way to physical strength, power will improve with repeated use. You will learn through practice and become more adept at making use of it to influence others.

Exploring the Principles

Here's a little exercise for you.

Think of a time recently when you were influenced to do something that you would not otherwise have done. Perhaps buying something that at first you were pretty reluctant to purchase. Or maybe, doing something for someone at work which you didn't really want to do.

Work through each of the principles and see how they applied to this situation. If you like, work through a number of different examples that you have experienced. It would also be helpful to apply these principles to someone else who you have witnessed being influenced.

What I'm hoping is that if you spend a little time thinking through your own examples, you will begin to see clearly how power is working. From this, it will then be more straightforward as the later chapters unfold for you to build this into your way of working.

Additional Concepts

Space does not allow for an in-depth consideration, and you probably haven't got time either. What follows are the bare essentials which will be expanded on a little later.

Power as Assets and Skills

When you have been thinking about the sources of power, you may have noticed that some of them, like money, qualifications, resources or position, create influence just by sitting there. Assuming people know you have lots of money, it will automatically begin to influence. Effectively, these types of power are working like assets that you own.

Others sources of power are only influential when you use them. Things like your network, your ability to build rapport or crack a joke, are generally only effective when exercised. In essence, these are working just like skills.

A source of power that is only effective when it is used is less efficient than a source of power that can work without being actively used. Why? Because you don't have to work so hard. Imagine, everyone knows you have a huge budget, and you have freedom to decide how it is allocated. People will automatically begin to work to influence you in their favour. They will come to you and be nice to you, offer you help and support and generally do whatever it takes to get into your good books. In return, they will do what you want them to do. That is much more efficient than continually having to persuade everyone that they should support you or do what you need them to do.

The higher you rise in the organisation, the more you need to be accumulating power sources that can work as assets. You don't have the time to communicate with every individual you need to influence. You may not even have the time to think about them. Senior people become powerful because of the assets they have at their disposal rather than their skills.

Thinking about power in this way yields a number of interesting and useful ideas.

- You can achieve influence without using your skills, provided that people know you have the assets and want a slice of them — or want to avoid them.

- The potency of different types of assets can vary depending on the principle of supply and demand.

- Many people fantasise and make assumptions about your assets, particularly if they are of the more obscure or intangible variety — like relationships.

- People without valuable assets have to use lots of skills to gain influence — the office politician perhaps?
- Assets can be divided between different people. Alternatively, they may choose to pool their resources.
- You can also lose your assets, spend them, invest them, or suddenly have their value drop like a stone.

Group/Organisational Power

You will probably have noticed that much of the phrasing above suggests personal ownership of assets or skills. This is true; however, the same also applies to groups of individuals, or even whole organisations. The combined effect of their individual assets and skills can become far more than the sum of their parts. This is what I usually refer to as group or organisational power.

Common sentiments about power include:

- "Marketing has the power to push that one through."
- "How come Sales always seem to get what they want?"

Although much of this is related to the combined individual assets and skills, that is not the whole story. To a large extent, the sort of assets and skills which are valued within an organisation are determined by the strategy the board or executives set in response to the external environment and shareholder expectations.

Put simply, if the directors recognise that market share is going to be squeezed, they'll turn the pressure up on the marketers to grab as much share as possible.

Therefore, all eyes turn to those who are capable of growing the business top-line — and budgets are usually made available. Alternatively, they may take the approach that they have to batten down the hatches and cut costs to protect profit margins — in which case, those who are in command of the numbers, financial experts, performance analysts and budget setters grow in power. This illustrates the point above about supply and demand.

In the context of the Stakeholder Influence Process, it is vitally important that you understand which of these powerful groups could have an interest in what you are doing (either positive or negative).

Formal and Informal Power

Another distinction that needs to be drawn is the difference between the power that is formally distributed around the organisation, and that which is acquired by individuals (very similar to personal power).

With a business, shareholders provide the executive team with a big bag of money, some ideas about the sort of business they want to create and an indication of what return on investment they need. The team then decide their strategy to deliver the return and set about organising and managing the business. They create the departments they need, allocate money to them and recruit people to head them up. They give these heads the authority to make things happen within their departments. Each budget is then subdivided again to create teams, and they then get on with the business. Of course, this is oversimplified, but it shows the basic principles of how the organisation formally starts to spread the power around the people and functions to get things moving.

Public sector and not-for-profit organisations operate in the same way, except that the key power asset driving the structure is likely to be the mandate and authority to act, with money being a secondary consideration (I'll sidestep the debate about whether or not this should be the case — that's just how it tends to work).

As the business matures, it will realise it needs to put in place some checks and balances. It, therefore, recruits a compliance manager and bestows upon him a remit to make sure that everyone is compliant, and if not, the power to force compliance. Again, this is a simple example of how power is formally created for the good of the business. Of course, this is not always as controlled by the top executives as they may like to think. I'm sure most people have witnessed "empire builders" collecting more and more departmental responsibilities. This is still an example of formal power.

However, this is very different from the informal power which usually takes the form of skills, capabilities and personalities. Experience and reputation also fall into this informal category. Curiously, qualifications are a formal asset usually regarded as an informal or personal asset, largely because it is given by a different organisation. Consequently, it has a high degree of independence from the current employer.

The reason for drawing your attention to this is that if you want to become powerful, remember that formal assets are generally given to people and/or can be taken away again; whereas informal assets are built and owned by the individuals and placed at the disposal of the organisation. The point to note is that if you want to build a solid power base, make sure that you build formal as well as informal assets as a sound investment in your future success.

As I move on through the book, I will return to these ideas about power and influence repeatedly because they explain how most things work within an organisational setting, and especially between people.

Key Points

- Power is the capacity to influence.

- The principles of power describe how influence works in the minds of those you wish to influence, and in your own mind too.

- It is impossible to avoid the subject of power if you want to succeed in organisational life.

- Power is responsible for all that is good in the world, as well as what is not so good.

- Personal power is the capacity of individuals to influence others.

- Group power is the capacity of groups of people to influence others.

- Formal power is generally bestowed on individuals. Informal power is generally acquired by individual effort.

- If you want to know why a particular decision went the way it did, look for the answer in the power dynamics between the decision makers.

Suggested Actions

- Reflect on what makes you powerful. What sources of power are working for you?

- Think about those who have the most influence over you. What is it about them that enables them to influence you?

- Pick up a copy of *Influential Leadership: A Leader's Guide to Getting Things Done*. This goes into much more depth on the subject of power and influence and many other useful topics too.

COFFEE BREAK

Emotional Project Managers

Mainly, project managers are known for their rational approach. Objectively tackling the problems and challenges which arise. Cool, calm and collected. They can maintain control of themselves and their process. There are exceptions to this, project managers who display high levels of emotion. They get passionate, enthusiastic and angry as they pursue their objectives.

Which is best?

- In what circumstances will the rational approach be most influential?
- When will the emotional approach be most influential?
- When does the rational approach struggle for influence?
- What about the emotional approach, when does that fail?

Personally, I believe that a balance needs to be struck between the two. In general, project managers are less influential than they could be because of their failure to utilise emotions.

Have you got the balance right? Could you make more use of emotion?

CHAPTER 5

Step 2: Identify

Now you are beginning to focus your mind on influence, particularly when taking the political perspective towards projects, there may be some significant changes to the people you need to be engaging with.

This chapter will help you to:

º   Challenge your thinking about who you should be engaging with as you seek to influence.

º   Identify the important stakeholders who can help you to achieve your objective, or stop you in your tracks.

º   Uncover stakeholders who you had not been aware of before.

If you're in a hurry:

º   If you are absolutely sure you know exactly who you need to be influencing and don't need any more challenge, jump down to the key points and move on. Otherwise, at least skim your way through this short chapter.

What is a Stakeholder?

The title *stakeholder* has been gaining popularity over the last few years. Its common usage refers to large groups of people, often with a somewhat nebulous, yet real existence — the media, the public and shareholders are often what's being referred to. As such, stakeholders used to be the chief concern of high-level boards and committees. However, a shift in popular usage is happening because of the realisation that the ideas used at a high level to manage these groups can apply just as well to more humble pursuits — the sort of projects that people throughout the organisation run.

The definition I want to use here is only marginally different from the one you would find in any dictionary:

> Stakeholder = Any individual or group who has an interest in your success.

A very important point to bear in mind with this definition is that "interest in your success" means that they will benefit when you are successful. It also means that they may lose if you are successful, or rather, they have a negative interest. If you are aiming for promotion to your boss' job, your colleagues may have similar aspirations. They have a negative interest in your success because if you get the job, they won't.

For example, you may be aiming to secure funding for your project which will deliver a new product for the Sales Director — who will be a positive stakeholder because they benefit from the delivery of the new product, assuming they want it. However, a colleague of yours may be pushing to launch a different product, and there is only so much budget to share around all the possible projects and products. If you get the budget, they may struggle. In this case, they are most likely to be a negative stakeholder — and someone who needs to be managed very carefully.

Incidentally, a stakeholder may not realise they are a stakeholder of what you are seeking to achieve, so some of them may need to be educated — but more on that later.

Another distinction is "individual or group." In the Stakeholder Influence Process, you will usually focus on individuals, but this largely depends on the scale of your goal and the number of people who are involved in the stakeholder management. Unless you are working on influencing via mass media, it is likely that at some stage you are going to need to focus on influencing individuals. I'll talk more about this at the end of the chapter.

Identifying Stakeholders

In this step of the process, you need to pull together a list of stakeholders — those who could help you or hinder you. This means people who can have an impact on what you are doing — and this is different from the group of people who are connected or involved in what you are doing.

Naturally, you can easily argue that anyone who is connected could have an impact, which is true, but not particularly helpful here. One of my favourite questions when I'm coaching with the Stakeholder Influence Process is who will benefit or win when you are successful? Quite often, the answer to this is everyone. This is good news; however, to be effective at stakeholder management, you need to get a little more focused while not completely ignoring the masses.

Instead, what you are looking for are those who can have the biggest impact. Those who could stop you dead in your tracks if they wanted to or, on the other side, carry you safely over the finishing line. Most people end up identifying eight to twelve impactful stakeholders — although these numbers often climb with subsequent reviews. If you struggle to come up with eight names, you may need to review the goal or objective you settled on in the focus chapter.

One of the biggest risks you have to protect yourself against is failing to manage the right stakeholders. I've found that generally people will manage the stakeholders who are most obvious, easiest to access and the ones they have the best relationship with. This is rarely enough to get the job done.

Stakeholder Categories

To stimulate your stakeholder identification process, here are a few different categories where powerful people could be hiding, waiting for you to engage with them.

Customers/Users

Perhaps a little obvious, but think carefully about this. Who is going to be using the result of your hard work? If you're implementing a new system, the people who will actually be using it are your end customers. They may be impactful themselves, but certainly their bosses will be able to make more of an impact on what you are doing. Similarly, if you're launching a new product, don't forget the people who will actually buy it at the end of the day. Often forgotten with internal projects, but with the rising power of social media their voice may be able to make more of an impact on your project.

Bosses

Again, a little obvious; however, less obvious are the bosses of the bosses; other senior level people elsewhere in the organisation who may be only indirectly connected with your project. Nevertheless, who could create quite an impact if they chose to — or you influenced them to!

Workers

These are the people who are doing the actual work within your project. They could be working directly for you, or they could be reporting to another stakeholder. The main thing is that these are the people with who you are likely to be in contact with frequently, and naturally they can have an impact on what is going on.

One caution is that it is easy to overestimate the level of impact they could have on the successful achievement of your goal. Of course, they could refuse to work — but the risk that poses is probably small, and the solution is likely to be a relatively easy one, unless they also fall into one of the other categories.

Advisers

These are usually very important people who sit alongside your project advising on all manner of things, such as legal, technical, etc. They could be very important, particularly if they are well thought of throughout the organisation. Sometimes, their advice to the board could kill your project.

Suppliers

Here I am referring to the people who provide their labour, skill or service to help you deliver your project.

They are generally outside of the organisation, yet can exert high levels of influence. Consultancy firms sit in this camp and their access to senior levels in your organisation make them very important stakeholders.

Movers and Shakers

You may think that this category is a bit of a wild card, but it is important to consider who these people are. They are the rising stars in the organisation. Those ambitious people whose power is rising along with their grade. They could be very important in helping you to really start to move fast. If they can be engaged and can see the benefit in what you are doing, they may be interested in getting involved. That they may also be interested in taking all the glory is also a consideration for you.

Disrupters

In addition to all of the other categories, there are a whole host of other people who could be affected by what you are aiming to achieve. It is from this group that you need to be ready for the "curved balls" and also where you can build powerful alliances. For example:

- Whose job will change as a result of your success?
- Who will not get the budget they were hoping for?
- Whose job will become more difficult?
- Easier?
- Will your success set a precedent that will make it easier (or harder) for others to follow?
- Who could be jealous of your success?

- Whose power will be disrupted because of the changes you are introducing?

Application to Your Goal

This is not a book that will just let you read. What I'd like you to do now is invest a little time in considering who your important stakeholders are. In the exercise below you will begin by identifying as many names as possible and then progressively refine them down to those who are the most important to what you are seeking to achieve.

Think about the goal you wish to focus on. Consider the categories above and assemble a list of potential stakeholders.

Now answer the questions below. Try to find new names to add to your list:

- Who can have the greatest impact on your success?
- Who are the key people who have influence over lots of other people on your list?
- Who are the biggest beneficiaries?
- Which powerful people could help dramatically if you were able to align them with your goals?
- Who's going to be really annoyed if you succeed and is capable of causing you lots of problems?

Homing in on the beneficiaries, add more people to your list:

- When you've succeeded, who is going to be better off?
- Whose life will be easier?

- Whose success will become easier when you've achieved your goal?
- Who are the end users/customers of your goal?
- Who will be able to achieve their own goals much quicker once you've achieved yours?
- Who else will be able to bask in your glory, share in your success?

And, don't forget those who are going to lose out when you succeed:

- Who is going to be worse off?
- Whose life will become more difficult?
- Who will have to change what they are doing because you've achieved your goal?
- Whose goal will be unachievable because you've hit your target?
- Who will not get the resources for their own project because everyone is working on yours?
- Who will be embarrassed by your success, perhaps because it shows they are lacking in some way?
- Who are the biggest losers when you are successful?

As you are going through this exercise, it is important to resist the temptation to restrict your list of names to those who you have access to. Stretch your ambition and note down the really powerful players in the organisation who, if on your side, would be able to make it happen. I'll talk later about how you may be able to engage with them. For now, just make sure to get these people on your list.

Review your list and find those who could have the greatest impact on your goal.

In many ways, this is a combination of the benefit (or loss) that they are aware of which could be heading their way and the amount of power/influence they have within the organisation (or at least in your stakeholder group). The more they are likely to win or lose — the more active they will be.

Review your list of names. Mark each one as high, medium or low in terms of the impact they could have on your goal. An estimate is sufficient: go with your gut feeling or get some help.

Hopefully, you have arrived at a list of eight to twelve people. Don't worry if you've got more. If you've got less, you might start to reflect on the usefulness of the goal you are focusing on. If you need to, return to the chapter on focus and review your thinking.

Working with Groups

If you have a very big goal on your hands that could affect a very large number of people, you should probably start off identifying the stakeholders by groups rather than individuals.

The reason for this is that you will find it very difficult to find the best people to be influencing. When focusing at a group level, you can use later steps in the Stakeholder Influence Process to work out which group (or groups) are most critical to your success. Once you've done that, you can settle on a subsidiary goal relating to each specific group. Then start identifying, analysing and mapping out the individuals. Let me give you an example.

A few years ago, I was working with a group of directors whose challenge was to influence their staff to get fully behind their change programme. On a flip chart, we compiled a list of impactful stakeholders such as suppliers, customers, employees and unions.

When we did the analysis and mapping (Step 3), they realised that the key to shifting the workforce was improving the relationship they had with the union stakeholder group — this was a revelation which had eluded them for months. So they settled on a strategy of focusing their efforts on shifting the position of the union (Step 4), which became a subsidiary influencing goal.

Three of the directors then huddled and considered who the powerful individuals within the union were; analysed them and mapped their positions. In that process, they discovered a number of small, but highly significant actions they could take.

Six months later, the HR Director defined this as the moment when the big turnaround for their organisation began. It ultimately helped them to avoid industrial action and build an extremely positive relationship with the union.

As I've said before, this process is not difficult. What makes it difficult is the confusion created by not using the Stakeholder Influence Process or something similar to view things at a higher level.

Many other chapters in this book can give you ideas on who the stakeholders with impact are; however, I'd like to add a small note of caution. Don't get lost in the process of identifying stakeholders. Once you've got enough to be working with, get on with it. You can come back later and refine your list, add some more or throw some off. Unless you start taking action, you're missing the point of the exercise.

Key Points

- Stakeholders are those who have an interest in your success, either positive or negative.

- Most people tend to focus on the stakeholders they know best, have the best relationship with, or are easiest to access. These may or may not be the best stakeholders to focus on.

- Powerful and prominent people not connected with your goal could potentially become stakeholders who can help.

Suggested Actions

- Review your new list of stakeholders for the goal you are working on with your line manager or mentor. See if they can add to the list.

- Get your team involved in this part of the process if appropriate. Many of these processes are easier, and quicker, when shared. It will also help them to raise their influence too.

- Keep your list handy, the next chapter may add new names to it.

CHAPTER 6

Developing Political Insight

It is one thing to accept that projects are set in a political context, and quite another to work out what that context is and what it means for the success of your project. What is necessary as a project manager is to develop a keen political insight so that you can judge how best to respond to it as an individual, and also how your project should respond and adapt.

This chapter will help you to:

- Learn a framework for analysis that can be used with any individual you need to understand more thoroughly.
- Analyse your key stakeholder's agendas.
- Build political theories about what agendas could be running behind the scenes.
- Root out faulty assumptions and find opportunities for greater collaboration.
- Prepare to compare and contrast agendas between people, including you.

If you're in a hurry:

- Skipping this chapter is a high risk unless you are already experienced and accomplished in political arenas. Your choice.

If you don't truly understand someone else's position, ambitions and problems, your influence attempts will be much more difficult. When others don't want to co-operate or are working against you, *conflicting agendas* or *politics* is often cited as the cause. This seems to signal a barrier to success, an explanation of why you can't get the things done — or is it just an excuse?

A more helpful attitude to adopt is that it is simply a question of different priorities. This opens up the possibility that the difference could be negotiated to achieve a *win-win* situation, while also removing the emotion (well, much of it at least). You are not in conflict with others — you've just got different priorities at the moment (well, you may be in actual conflict, but often it's a figment of your imagination).

Most of the difficulties arise because of lack of awareness of what the other person has on their list of priorities. Unless you know what they want, how can you negotiate? Until you know what they are trying to *win,* you cannot come up with an innovative idea that can enable you both to *win*. Instead, you are just guessing — or more likely, just pitching your ideas and hoping they'll be accepted.

To overcome this drag on your performance, you need to carefully assess the agenda of those who are important to your success by developing political insight.

Types of Agenda

Agendas come in two main types, professional and personal. The professional agenda is all the work-related priorities: performance targets, job descriptions, project plans, etc. Some are often visible or easily revealed with a question or two, so they are easy to work with. More difficult to spot are the subtle influences on the professional agenda. A profit warning can put unseen pressures on key people in the organisation, which may not be openly talked about. Sometimes, the way of teasing this out is to look for the drivers behind the public or professional agenda. You need to dig deeper and look around more corners to see what is really making things happen.

Personal agendas are much more difficult to work out. Items here include career goals, bonus aspirations, or even settling old scores and getting revenge. Without a good relationship with the individual stakeholder, a high degree of intelligent guessing is required to work out the personal agenda. It may mean you have to seek insights from people in your network and consider recent history and behaviour patterns. However, any attention you put towards uncovering the personal agenda will yield big results because for many people the personal agenda is their key driver. Of course, they will never admit to being driven by personal gain or greed — but there is a bit of that in all of us, isn't there?

The other type of agenda that is often cited is the hidden one. This suggests a clandestine motive and scurrilous activities. That's one way of looking at it. The other way is that it is simply an agenda they haven't shared with you, for whatever reason. It doesn't matter whether it is the personal or professional agenda that is beyond your awareness, the gap still represents a risk that needs to be managed.

Rather than label agendas as hidden, I much prefer to think of them as simply unknown. Your job is to uncover what is really going on, and the next section is going to really test your knowledge and in all likelihood, expose significant gaps in your awareness.

Exploring Political Agendas

In order to gain political insight, you need to be exploring an individual's political agenda. This term is really just an over-arching label for the combined policies or issues that an individual has decided to further, or resolve. You could almost regard this as a personal strategy or manifesto to satisfy their needs and ambitions.

Apart from at governmental level, or in party politics, political agendas are rarely explicitly stated or even understood by the individual. Consequently, a little investigation work is needed. In the same way that it is unwise voting for a political candidate on a single issue, so too it is unwise to make decisions about what is motivating someone, or what they may do next, on just a few clues.

You'll have to judge the depth you need to go into, and this section is going to really test your understanding while also providing you with a checklist of things to think about any time you want to do some political analysis.

As you read the following sections, I'd like you to keep in mind a specific and important stakeholder related to your goal. Preferably, someone you know reasonably well. As with any individual you analyse, the aim of your enquiry is to:

1. Explain their current behaviour.

2. Understand their influences.

3. Predict how they may respond to your influence.

Below are four key areas that, in combination, will help you to understand an individual's political agenda. You will quickly realise this is extremely complex, and I fully accept that you are unlikely to be able to answer all of the questions I pose, even if you know a stakeholder really well. That's normal. My job is to chart the territory and offer you the assurance that any effort given to increasing your understanding will lift the probability that your predictions will be accurate.

As you read, reflect on how you would answer these questions for the stakeholder you have in mind. The exercises that follow this section will pull together your thinking, help you to prioritise the gaps, and enable you to begin drawing conclusions.

Career

What do you know about their career? Reflecting on where they have come from, and where they are going can tell you a great deal about how they behave, or may behave, in different situations. These only provide generalised clues, yet they can be surprisingly accurate. Exploring this area can also shed light on who they have an affinity with, their values and also their motivations.

Some key questions you need to be able to answer include:

- Where have they worked in the past? Consider companies, countries, divisions, industries, professions, functions, etc.
- What was their last position all about?
- Why did they leave their last job?

- Who recruited them into their current position?
- How long have they been in their job?
- What external positions have they/do they hold?
- What are their career ambitions?
- What will their next job be, and when will they get it?

Starting to sound like an interview? That is no coincidence, because recruiters want to understand what makes them tick, and what performance they can expect if they appoint them.

Performance

People behave and respond differently under pressure. Role or job performance for many is an acute source of pressure, especially when it is not going so well. If it is going well, they'll be far more relaxed about things.

The important thing to bear in mind is not what you think about their performance, but what their bosses think. You may well think they are rubbish at their job, but if they are pleasing their paymaster, they will not be under as much pressure.

Without access to their performance reviews, you're going to need to consider other clues, such as:

- What is their formal job description or terms of reference?
- How do they appear to be performing against these?
- To what extent are they calm and confident?
- When do they become more agitated?

- How are they interacting with their boss?
- If they are not performing well, why?

It also needs to be recognised that not everyone is concerned about their formal performance. People who are extremely confident, or personally secure, may not worry terribly what their line manager thinks.

Friends, Associates and Enemies

Here you need to think about all the people around the individual you are analysing. You can consider these to be personal stakeholders with a personal interest in the individual. The interest could vary widely from enjoying their friendship, love or hate. Alternatively, these connections could be forged on the basis of shared values, provenance, clubs or even family ties.

It is important to identify who these people are, the nature of the relationship, and finally, their impact on the work they do.

Questions that begin to illuminate this include:

- Who are their friends? Who do they dislike?
- Whose company do they seek out before/after meetings?
- Where do they socialise?
- Who are their most important network connections?
- Who do they rave about? Who are they openly critical of?
- What is the nature of these relationships?

When thinking about the potential impact of a relationship on the individual, watch out for the way that their personal relationship affects their professional relationship. Some people are very clear on the boundaries and are well able to be hard as nails in the boardroom, yet still go out for a social evening afterwards.

Power and Influence

What makes this individual powerful? Refer back to the relevant chapter and get specific about the sources of power which they use to influence others. For each source of power, try to pin down exactly what it is about that source that helps them.

For example, if you think they gain power from their credibility, what exactly is it that makes them credible to those around them. Is it their qualifications, track record or intellect? Or, is it something else that makes them credible in the eyes of those they influence?

Also, think about who they can influence and who they fail to influence. How strong is their influence with different people?

Finally, consider influence going in the other direction. Who influences them and why? What sources of power are these people using to influence the person you are analysing?

Additional Factors

There are many other areas that you may need to consider. Rather than test your patience, here is a summary of the key ones that occur to me immediately:

- **Values.** What things do they consider to be important? What you're looking for is the most important of those. For instance, do they place integrity above all else? What about process efficiency, cost-saving, relationships, competitiveness? What are the top three values they exhibit?

- **Behavioural Style.** Closely linked to personality, the way they tend to behave in relationships. I will go into this in more detail in a later chapter. Right now, think about what patterns you notice in the way they behave. For instance, do they tend to be highly sociable, or prefer to concentrate on their work? Are they serious or jovial?

- **Motivations.** What really gets them excited? Either positive or negative, if you notice their emotions rising on particular topics or when around particular people, it may give insights into their motivations, values and also their agenda.

- **Cultural Heritage.** The culture they were born in, brought up in, have lived in, and worked in, all contribute to their values and behaviours. Learning to notice how each culture is influencing their behaviour can give you vital clues about the way they will react in difference situations.

There seems to be an unlimited range of things you could think about when doing your political analysis; however, you need to judge the relative value. In a moment, I will move you on to the next stage, but I cannot leave this section without challenging your knowledge about your stakeholder one more time. Here is a set of questions that are deliberately posed to jump your thinking off the tracks and help you conclude with a rounded view of what might be going on.

Try to answer each question and also consider why that is the case, or at least, what makes you think it is:

- What things do they tend to focus their attention on?

- Who is their favourite person?
- What issues/problems are they avoiding?
- What scares them?
- Who is putting them under pressure?
- Why might they be happy to retire tomorrow?
- What is exciting them the most right now?
- What are their personal goals and ambitions?
- What problems are they struggling with right now?
- What are they proud of?
- What do they enjoy doing the most?

Distilling Political Insight

Hopefully, as you have read the preceding section, you have been reflecting on one individual stakeholder who you know reasonable well. If you haven't done that, please go back and do it now because I have a little exercise I'd like you to complete.

A notebook is necessary for this because I'd like you to work through the questions below and record your answers. Try to answer each question succinctly with as few words as possible.

- What is their professional agenda?
- What is their personal agenda?
- Can you list their top three priorities?
- Their three most important stakeholders?

- How about three things that are keeping them awake?

- Finally, can you summarise their political agenda?

I wonder how you got on. Since this is supposed to be someone you know quite well, I would like to think that you did quite well with this. If you did, great, but I bet there is a fair amount of uncertainty in your answers. Hopefully, not too many enormous gaps.

Filling in the Gaps

Regardless of how well you were able to complete the exercise above, there will be gaps in your knowledge. Developing political insight in an ongoing activity. The more times you come back to it, the deeper your insight will be. Ideally you need to make the acquisition and analysis of political agendas a habit.

To be honest, it can become absorbing once you start to get to grips with what you need to look for. Most people shy away from it because they are starting from a very low level of insight and skill to acquire it. Yet, you only have so much time to give to this activity, and you need to move fast. Some prioritising is in order.

Firstly, keeping to the same stakeholder you have been working on:

- Where are the biggest gaps when you completed the exercise above?

- What are the key things which you don't know that you need to know?

- What assumptions are you making about them?

- What are the critical things to be accurate on?

Okay, now go fill the gaps.

Tempting as it would be to suggest you should just walk up to them and enquire about their political agenda, that's unrealistic in 99% of cases. That would be a little crass and unlikely to get a clear answer. Either because they are not clear what a political agenda is, what theirs is or they are very suspicious about why you're asking.

Instead, build your routes to insight from these ideas:

- **Ask a friend.** Swapping notes with trusted colleagues. You can vary the intensity from an informal chat about the world over coffee to a full-blown joint assessment. Picking up little clues here and there is the bread and butter of the intelligencer.

- **Surveillance.** Watch them closely with your questions in mind. See if you can collect more evidence of their motives and influences. This is a good thing to be doing anyway. Always keep one eye looking for more political evidence.

- **Guess.** Yes, that's right, go on, have a guess. You may be surprised how close you will be. Once you've guessed, do some detective work to find out if you are on the right lines.

- **Pretend.** This is not a technique for everyone — but try it before you dismiss it. During a quiet moment when nobody is looking, sit in your chair the way your stakeholder might sit. Imagine you are your stakeholder. What might he/she be thinking about their work? What problems might be playing on their mind? How might they react to different people and why? Go on, give it a go.

- **Ask them.** Not directly, but take every opportunity to get to know them. Build trust and see how quickly they begin to open out and tell you more about what is really going on.

Personally I am a little bemused by how reluctant people are to make use of idea five above. Assuming you have a reasonable degree of trust, and a modicum of inter-personal skill, this is probably the easiest and most reliable way of getting to know someone's agenda. And, while you're having a good chat about the world, they are getting to know you too and, there is a distinct possibility that you may be able to overcome so many obstacles in the process.

Not convinced? Okay, carry on guessing and running the risk of making dangerous assumptions. Your choice.

Making Comparisons

The next step in the Stakeholder Influence Process is to build a picture by analysing all of the important stakeholders that you have identified. To its fullest extent, this may mean you have to explore all of their political agendas in the way you have done in this chapter. If you do, I can promise you will benefit, big time.

Understanding one stakeholder's agenda may be interesting but not terribly helpful unless you only want to influence one person. Assuming that the goal you are focusing the Stakeholder Influence Process on is of reasonable size, you are going to need to influence many stakeholders, and all of them will be connected with each other in some way.

Effective stakeholder influence is about carefully observing how other people's agendas match or mismatch with yours. However, it is also about observing how their agendas are competing with each other. Decisions are always about one agenda meeting another agenda and somehow reaching an agreement. This makes for a much more complex picture.

Most people only think about their own agenda and then look for ways to get others to agree with it. To be really successful at influencing someone else, you need to fully understand their agenda and then bridge the gaps sufficient for them to agree. At an exceptional level, you will be able to manage and negotiate between all the agendas that have a vested interest in what you are doing. That's real stakeholder management.

Key Points

- Effective influence requires a good understanding of the other side's agenda.

- Faulty assumptions sit behind most influencing failures.

- Investing time in developing insight will give immediate returns and increasing benefits as the years roll by.

- Personal agendas are more influential than professional ones for most people, even if they don't admit it.

- The easiest way to uncover personal agendas is to build high levels of trust.

Suggested Actions

- Draw up your own political agenda. Use the ideas in this chapter to analyse yourself.

- Complete a political analysis on your five most important stakeholders.

COFFEE BREAK

Translation Issues

Most project managers are speaking the wrong language. What they do is talk in the jargon and terminology of their profession, and why not? It helps them to communicate with each other far more effectively. Because each project manager understands the complex ideas conveyed with simple words, they can economise on the words they use. Trouble is, people in the business are doing just the same, but they are using a different language.

If you want to be understood as a project manager, you need to understand the language of your stakeholder. You need to go beyond mere conceptual understanding. You need to really understand what they are talking about. That is not easy to do, but it is vital if you want to become influential.

Once you've done that, you can begin to translate your own language into words and phrases which connect directly with their world, in the way they would express it.

- ○ What key words and phrases do you and your colleagues love to use?

- What key words and phrases do your key stakeholders love to use?
- What expressions might your stakeholders struggle to understand?
- How can you check your understanding of what they are talking about?

Don't think that because they fail to understand your terms they are stupid. They are not, and nor are you. So, work hard to root out the potential for misunderstanding.

CHAPTER 7

Step 3: Analyse

You have probably had enough of the difficult questions I have been posing for you so far. If you have developed a reasonable political insight around your project and its stakeholders, now is the time to bring the pieces together and start to work out what to do in order to make greater progress with your project.

This chapter will help you to:

- Consider the degree to which your agenda matches or mismatches with each stakeholder.

- Reflect on the quality of your relationship with each stakeholder.

- Build a complete picture of the stakeholders surrounding your goal or project.

- Begin to identify themes and patterns in your stakeholder community.

- Start to identify actions you can take to improve your prospects.

If you're in a hurry:

- This chapter is central to the whole process; however, you can skim it and probably get moving quickly.
- Make sure and draw a map of the stakeholders around your goal. The map here is likely to be different from anything you've done before.
- Then keep coming back here to develop your practice.

So far, I have been sharing a lot of information about what you need to know, especially about stakeholders and their agendas. I hope you have been applying it to something that you need to deliver, because now it is time to bring them all together.

Combining your insights into a stakeholder map is vital if you are to be able to create a successful strategy to achieve your objective. Chances are high that you have far more stakeholders than you could possibly engage with, so there needs to be a way of safely prioritising what needs to be done.

There are many different approaches to stakeholder mapping, and the one I am about to introduce you to is unusual. I have been evolving this over the last ten years, and it is proving to be extremely successful, and easy to use. What it does differently is place the quality of your relationship with each stakeholder centre stage.

The Stakeholder Map

Take a look at the map below. Two things you will immediately notice are:

- One of the dimensions is titled "Relationship." This is because, in my view, the quality of your relationship with each stakeholder has a dramatic effect on your ability to influence them. By focusing your attention on building the relationship, you will enhance trust, increase disclosure and maximise influence.

- The titles in each of the boxes are provocative. I will explain the rationale behind each of these later, but right now I want to stress that the purpose of using these labels is to stimulate your thinking, rather than label your stakeholders as *enemies* or *players*. They may well be, but it is usually unwise to tell them that. So keep this to yourself.

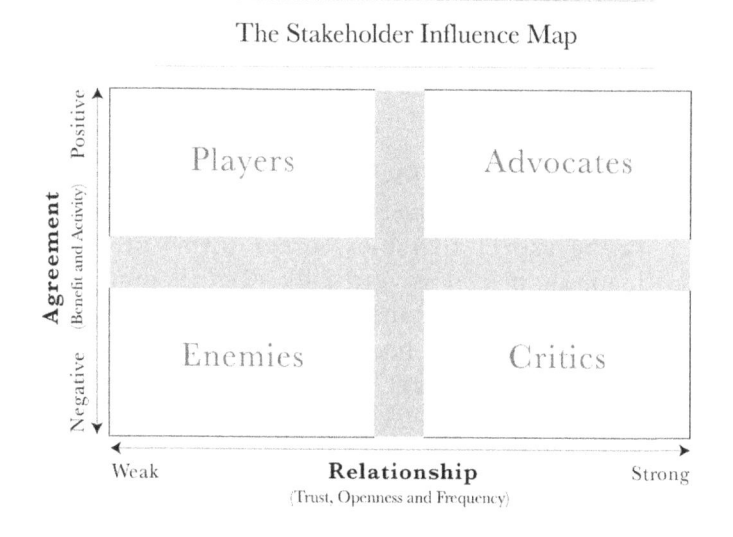

The Stakeholder Influence Map

	Relationship (Trust, Openness and Frequency)	
Agreement Positive (Benefit and Activity)	Players	Advocates
Negative	Enemies	Critics
	Weak → Strong	

The general idea is that you select the eight to twelve most powerful stakeholders who can help or hinder progress towards your goal, and write their names (or initials) somewhere on the map. It is important to focus on a single goal each time you use the map. Most people I coach generally have 4-5 maps on the go at any time because they have several important things they are aiming to influence within their role.

For each individual, you need to make a quick assessment of where to position them on the map. It is not critical that you have precise answers, sometimes gut feeling is all you can go on. The more you use this technique, the quicker it will become. When you keep it under review, initial assessments will become more accurate as you learn more about their agendas, or begin to make progress with them.

Another important feature of the map is the grey spaces between the boxes. The purpose of these is to accommodate people you are unsure about. In most cases, when first mapping out the stakeholders around a goal, there is a great deal of uncertainty about the position. This is usually because the Stakeholder Influence Process is quite good at identifying stakeholders you haven't considered before, or even met.

Before I move on to describe how to determine where on the map to place stakeholders, I'd just like to touch on an aspect that most other forms of stakeholder mapping prioritise. Virtually all of them use a dimension of power or influence in place of relationship. This is important because what they are doing is recognising the impact that the stakeholder could have on the goal.

However, in my approach, I place more emphasis on relationship because it is more useful when it comes to determining an effective influencing strategy to land your goal. My way of handling power and influence is simply to only consider stakeholders who have the power and influence to affect your progress.

Incidentally, another dimension often used by other approaches is *interest*. To me, this is synonymous with *agreement*, and I have chosen this word because it conveys a sense of action and practicality. As you will see later, *interest* is fully catered for.

The Relationship Position

To decide where to place a stakeholder on the map in terms of relationship, you need to think about the quality of the relationship you have with that individual. The three main areas to think about are:

- **Trust.** This is the most important and seeks to determine the extent to which trust exists within the relationship you have with the individual. This is a two-way consideration — do you trust them, and do they trust you? Make sure to consider the evidence. Many people I work with recognise their feelings about trust but are unable to quantify why. They are also prone to distrust people based on hearsay or events that happened years ago.

- **Openness.** This reflects the way each party volunteers information. It is one thing to have trust, but this does not necessarily mean that they will go out of their way to warn you when they learn something that could be helpful to you. Equally, are you proactive in lending them a hand or tipping them off about something that could be useful to them?

- **Frequency.** Both trust and openness can exist in a relationship, but if you haven't seen them for years, the relationship is not going to be terribly helpful for you. If you are regularly chatting with them or dropping by for a coffee, chances are high that the relationship is of the strongest kind.

The stronger each of these elements is in your relationship with the stakeholder, the further to the right you will write their name. Likewise, if these elements are missing or patchy, their name is going to end up heading left. If you're not sure, perhaps because of conflicting evidence, or you don't even have a relationship with them, their name is going to land somewhere in the grey area of the map.

The Agreement Position

To finish the positioning of your stakeholder, you also need to consider the degree of positive or negative agreement the individual has with what you are aiming to achieve.

This is one reason why it is so important to clarify exactly what you want to happen. If you're not clear, they won't be clear, and it will be impossible to position them on the map. This is also why you should use a different map for each goal you are working on. Some stakeholders will appear on more than one map and could be in a different position based on agreement.

There are three important elements to consider when it comes to agreement:

- **Interest.** What benefit will they get when you have successfully achieved your goal? If it will help them to solve some of their problems, or make them money, they are going to fall nicely into the top half of the map. Alternatively, if you are going to make their life more difficult, or maybe even jeopardise their job or promotion prospects, they will be heading for the bottom half (interest can be negative as well as positive).

- **Agreement.** Do they agree with what you are trying to achieve? If they think that your goal should be achieved, even if they don't personally benefit, then they are likely to be quite helpful, and you'll be placing them in the top half. Similarly, they may be in line for benefits, but perhaps they can see a wider and negative impact on the overall organisation, which could mean they disagree. Or perhaps they can see even bigger benefits arising from someone else's project!

- **Activity.** Are they actively supporting you, helping you to clear through the issues and roadblocks? Activity is often an indicator of their agreement and perceived benefit. Of course, they may be very active trying to stop you in your tracks, but not for long I hope.

It is on the Agreement dimension that you will start to gain real progress if you have been able to be very focused with your goal. If it is crystal clear what you are aiming to achieve, it is much easier to determine if an individual is in agreement with you. Even so, if this clarity is new, you may not be able to position the stakeholder if you haven't had a chance to ask them yet. In this case, the stakeholders will initially land in the grey area between the boxes. It is important to remove uncertainty as soon as possible, particularly with highly impactful stakeholders.

So, if there is firm evidence that an individual is agreeing that your goal should be achieved, that there are benefits for them and they are actively working on your behalf, write their name in the upper half of the map level with where you have assessed the relationship. Evidence to the contrary means they are likely to land in the bottom half. Similarly, if there is mixed evidence, they'll be in the grey zone.

Prioritising Your Stakeholders

As mentioned earlier, most people have far too many stakeholders and need some way of focusing only on the most important eight to twelve. Doing this will help you to:

- Spend more time with those who matter most.
- Avoid you being distracted by loud but largely unimportant stakeholders.
- Give you more time to prepare adequately for influencing the right stakeholders.

There are occasions when it is necessary to expand the number of stakeholders you are engaging with, for instance:

- When you are pushing forward a large and complex goal. In this instance, you might start with your big goal, assemble your stakeholder map and then define smaller goals where you can get a little more focused.
- When you are using this approach to stakeholder mapping with your team. Sharing the task of influencing stakeholders makes it feasible to incorporate more. However, make sure that you are certain that they are still worth engaging.

A good way to begin prioritising your stakeholders is by analysing each in terms of their degree of interest and awareness.

Take a look at the diagram below. The idea here is to transfer your long list of stakeholders (hopefully you developed this in an earlier chapter) into the columns. Base your positioning on your opinion of the extent to which they will gain or lose if your goal is achieved, and the extent to which you believe that they are aware of this.

It is quite common to find that some of the major beneficiaries of a project or initiative simply don't know what it could do for them. Some don't even know that something is happening. Decisions made in committees and at senior levels are seldom communicated effectively. Even if they do know what you're doing, they may not have noticed (or had the time) to connect it with their agenda.

Stakeholder Identification

	Big Winners	Winners	Marginal Win	Marginal Loss	Losers	Big Losers
They know						
Unsure						
They don't know						

Here's a little exercise to get this working for you.

Take a page in your notebook and draw out the table.

Keeping your mind on a specific goal you are working on, fill the table with the names of as many stakeholders as you can think of.

Circle the names of those who you consider to be powerful and influential in relation to your goal.

Now decide which stakeholders you need to focus on. Bear in mind:

- Prioritise those who are powerful and are big winners or losers.

- Unaware, powerful, big winners are huge opportunities as they may swing into action immediately.

- Unaware, powerful, big losers are significant risks so handle with care.
- Are their opportunities to increase the win for powerful marginals?
- What can you do to lessen the loss for powerful losers?

The purpose of this exercise is just to start you thinking about who to focus on rather than come up with specific actions — that will come in the next chapter as you start to develop a clear strategy.

Assuming you have settled on a list of eight to twelve stakeholders, you are now ready to complete your map.

Completing the Stakeholder Map

Draw out the map in your notebook, including the grey areas, title it and position each of your stakeholders.

As you are doing this, remember:

- Remain focused on their agreement with your chosen focus/goal. To what extent do they agree that your goal should be accomplished?
- Someone may agree with you that your goal should be achieved; however, if you have specified that it should be done by the end of this year, they may disagree with that. If so, they need to be placed somewhere in the bottom half.
- Go with your gut feel. One way of looking at this is that you are creating a hypothesis or theory that you will progressively test and improve.

- Be brutally honest. There is no benefit to be gained from over-estimating the strength of agreement or relationship.

Here are some more ideas about the categorisations:

- **Advocates** are people who you have a great relationship with, are on your side and really want you to succeed. You are likely to be close to them. Line managers often go here. People who are merely supporting you are likely to be entered towards the middle of the page, while those who are shouting from the rooftops for you will head towards the upper right.

- **Critics** are individuals who you have a good, open relationship with, but are perhaps your devil's advocate. They can see the flaws in what you are trying to achieve. They are great at pressure testing your goals and plans. If they are really active in trying to block you then they will probably be right down towards the bottom of the box.

- **Players** are people who will agree to do something in a meeting, but then fail to follow through. They say *yes* and do *no*. Often the problem is that they are not being straight with you, or perhaps their intention was good, but after the meeting someone else influenced them in a different direction.

- **Enemies** are clearly against you achieving your goal, but you can't quite work out what they will do next to stop you. Loose cannons is another phrase that springs to mind. Most of the time people land here for the wrong reasons. Lack of engagement means you don't know them that well so the scope for misunderstanding their motivations can be large.

As you are considering each individual, challenge yourself with these questions:

- What is their professional agenda?
- What is their personal agenda?
- What will they gain if you succeed in reaching your goal?
- What will they lose if you succeed?
- What will they gain if you fail to achieve your goal?
- What will they lose if you fail?
- How does their agenda compare to yours?
- Why have you placed them in that position on the map?
- What evidence is there for the level of trust?
- What evidence for the level of agreement?
- How are they connected to others on your map?
- What makes them powerful?
- How are they opposing/supporting you at present?
- What political considerations are there with this individual?

Drawing Conclusions

It is tempting to jump to answers and actions. You might be right, or you might be wrong. Pause a moment longer and use these questions to draw your thoughts together about the whole stakeholder map.

- What are the main themes around agreement and disagreement among your stakeholders?
- Are there other organisational influences holding them back?
- How are the people on your map connected (or disconnected)?
- What gaps are there in your knowledge/insight that need to be filled?
- Who can help you learn more?
- Is there anybody else you have forgotten to consider?

If you're thinking that one of your stakeholders could fit into two boxes — they can't. They must fit into one box, or between the boxes. If you feel the need to place them in two boxes, it may be a lack of clarity on what you are trying to influence — your goal specification.

Because of your goal, you may have been mapping groups rather than individuals. At some stage, you will need to generate an additional focus (and a new map) for each specific group you want to influence. That way you can break up the group and look for the opportunities to influence individuals within each group.

A final point to remember is that the purpose of this part of the process is to quickly get you thinking about practical actions you can take in order to move things forward. Often, when I'm challenging someone with their map in front of them, they discover critical new ideas that can accelerate their progress within ten minutes.

While the immediate ideas are really useful, they are generally more tactical in nature. For most goals of a reasonable size it is worth investing a little more time to develop a strategic approach. To this end, over the next few chapters, I am going to deepen your understanding of the environment where you are operating. I will explore the different types of stakeholder and some alternative approaches to them that may be appropriate to your situation. This will contribute greatly to the development of your influencing strategy.

Key Points

- The purpose of this approach to stakeholder mapping is to capitalise on the importance of the relationships and quickly move towards practical action.

- Use one map for each goal you are working on. While the stakeholders may remain the same, their positions will probably vary from map to map.

- Don't leave maps lying around where they can be seen by people who don't understand how you are using them for positive and constructive purposes.

- Trust is of critical importance and often the majority of the action is aimed at improving the position of stakeholders on the relationship dimension.

- Don't agonise over your assessment here. If you find yourself struggling to put pen to paper, do it quickly, take some action and review it again soon.

Suggested Actions

- Make sure and complete the map for at least one significant goal before you move on to the next chapter.

- In fact, do a map for several of your key goals.
- Share your thinking with either your line manager, a trusted colleague, or both.
- Make sure and review each map regularly.

COFFEE BREAK

Hopeless Projects

When you do the analysis, on occasion you may find that the harsh reality is that you are not going to be able to deliver on your project. Either because the resources cannot be made available or there are powerful people who will never allow it to happen. When this happens you need to consider:

- Often it is better to face this reality and deal with it rather than stick your head in the sand.

- Developing your capability to influence will place you in a far better position to challenge what is going on.

- Challenging the continued feasibility of your project takes courage and conviction.

- Proactive stakeholder engagement is often the only way of bringing these situations to a good conclusion.

- Would you rather be working on a hopeless project or seeking out new exciting opportunities?

- Proceed with lots of caution and also, regular discussions with your mentor.

My personal view is that competent project managers who have the courage to bring about a premature end to their projects are the most valuable of project managers. It indicates people who are capable of rising above their personal agenda in order to deliver outstanding value to their organisation.

CHAPTER 8

Understanding the Bigger Picture

While it may seem a little odd to be considering the bigger picture after you have done your analysis, I find that most of the time the bigger picture isn't capable of being seen until you have analysed your stakeholders. Once you have considered the ideas here, you may need to revisit your earlier analysis before beginning to finalise your strategy and plan.

This chapter will help you to:

- Appreciate the wider environment as the context for your role.

- See how the environment may be shaping the progress of your goal.

- Identify the risks that need to be handled.

- Recognise new opportunities to link to other agendas and accelerate your progress.

If you're in a hurry:

- You may well be rushing in the wrong direction unless you've got this one nailed.
- And, if you can nail this one, you'll be able to move even faster with greater surety.
- Up to you if you skip this one.

If you only think about each stakeholder as an individual in your analysis, you will miss the way they link together in the bigger picture. Quite possibly, you will also waste a great deal of time and effort. The only way you can develop an effective influencing strategy is to consider all of your key stakeholder's interests together.

When you do this, you may find that you don't have to please them all. In fact, it may be prudent to upset a few to advance your strategy. That's why being able to see the big picture will help you perform.

In this chapter, I want to challenge you to consider three things that could help to shape the overall direction you need to take:

- **Political Scenarios**. All organisations are political by nature. They are inhabited by people with agendas and ambitions. The external environment is putting the organisation and its top people, under pressure to change or die. These factors give shape to the political structure of the organisation, and the way this plays out could have a far-reaching impact on your goal. In strategy formulation, it is vital to understand the political context for what you are doing.
- **Risk Identification**. At a slightly more tangible level, the vast array of change that is taking place within your organisation could have a major

impact on what you are doing. Most people tend to think of the immediate risks to their ability to implement rather than those a little further out. Keeping sight of all major risks and factoring this into your strategy will help to make it more robust and more successful.

- **Opportunity Identification.** Similar to risks that have a negative impact, opportunities could have a massive positive impact on what you are doing. Despite this, few people stop to consider what else is going on in the organisation that could give them and their goal a big boost. Once spotted, incorporating them into your strategy could help you to foster more buy-in and attract more power to your purpose.

When you've gone through this chapter and tried out the exercises, I hope that you will then have a much greater appreciation for the position of each of your individual stakeholders, and will be ready to finalise your strategy.

Organisational Context

Before you can compile your scenarios, risks and opportunities, you need to take a good look at the organisation you are working in. The challenge that this presents to you will vary depending on your current knowledge and experience. The more you know, the better you will fare. The less you know, the more work you will have to do, or, the higher your risk of getting it completely wrong.

As with many elements of this book, the amount of attention you pay to the content is up to you. You will also need to decide on a definition or scope for your assessment. Depending on your work and your goal, you may choose to focus on your industry, country, market, company, division or function. Bigger goals naturally need higher levels of analysis, and a greater investment of time will be necessary.

Right now, my advice to you is to do this quite quickly based on what feels appropriate. As you go through the various sections below, pause and make notes that you can return to later as you go deeper. What is likely to happen, especially the first time you do this, is that a very complicated picture will emerge, and you may struggle to make sense of it all. Each time you do this, more will come into focus, and this is a natural part of the learning process.

So, persist with these investigations and you will profit enormously.

Purpose

What is the purpose of the organisation you are analysing? What job is it meant to do? Why was it created? Your initial thoughts will probably relate to the formal or public objectives, and these need to be checked out with reliable sources. You'll also need to delve a little deeper and try to establish the hidden drivers too.

Challenges

In the pursuit of its objectives, what's actually happening right now? What challenges and issues is the organisation facing? What is getting in the way and making it think really hard? This might be external competition or internal process problems. Put another way, what is getting in the way of it achieving its purpose?

Strategy

Given the purpose and challenges, what strategy is the organisation adopting? Again, this will take some time to figure out, and the best way of doing this is by talking to the people who are in there doing it. Building high-trust relationships in the main parts of the organisation is vitally important if you need to understand what is really going on.

Structure

As you probe into the areas above, you'll be deepening your understanding of the structure of the organisation you are looking at. Not only do you need to know the formal structure, but also the intricate web of informal groups and alliances that enrich (or complicate) the picture. In many places, it is these informal groupings which drive the real action.

Powerful People

All structures are full of powerful people. People who are making a greater impact than those around them. As your awareness grows, pause and think about who these people are. Remember the sources of power and look far and wide for people who can influence those around them. In addition to the usual powerful positions, also consider the talented and ambitious who are on their way up the ladder.

Politics

What political agendas are the most powerful people running with? Refer back to the chapter on political insight and invest some time investigating these people, especially those closest to what you are doing in your role. As you pull together these insights, notice how their agendas match and mismatch with each other. This is how you can begin to build your political theory of the organisation and will help you to understand the real undercurrent to the decision-making processes.

Groups

An earlier topic to consider was the organisational structure. As you did this, you will have identified many different groups. Moving closer to your goal, apply each of the sections above to understanding each of the groups that have a potential impact on what you are doing. This is a lower yet more relevant level of analysis that will help you build a concrete strategy.

Remember to analyse informal groups too. These can be difficult to spot yet hugely influential in the cut and thrust of organisational life.

The analysis suggested here should become an ongoing part of your work, continuing to build your awareness and understanding of what is going on around you. As your insight grows, so will your confidence that you are selecting the right strategy to move your goal forwards. You will also notice that you will begin to relax and have more fun.

Developing Political Scenarios

Based on your thinking above, now it should be possible to refine your political theory of the organisation into a number of scenarios, or outcomes that may happen. When you do this, you will really start to crystallise the big picture and be able to recognise your part within it. This makes selecting the correct strategy far more reliable.

Here is a basic process you can use to develop scenarios to inform your strategy development.

Assemble the Options

This is far simpler than it may appear to be at first. If you have invested a reasonable amount of time thinking about the topics above, you should be able to answer this question:

Over the next few months (or years) what big changes could impact the organisation?

These changes could be internal or external. They could be positive or negative. Here are some ideas to illustrate what I mean:

- The business will fragment into a number of Strategic Business Units.
- The Finance Director will be promoted to CEO.
- The Centre of Excellence programme will be abandoned.
- One of the two key products will be split from the main company and sold to a competitor.
- Nothing will change in the next two years (always a possibility).
- Finance will lose control of IT as a CIO is appointed.

Spend some time brainstorming all the things you think could happen. You may wish to do this over a few days and get some input from other friends and colleagues in your network. What do they think might change in the future?

Select Your Scenarios

Since you are preparing to build an influencing strategy and plan to achieve a specific goal, my suggestion here would be to review your list and pull out two or three which would have a major impact on what you are doing.

Don't throw away your bigger list. You can save these to review again later in case they become more relevant. It may also be that you don't currently realise the significance of some of these potential changes. Again, a good idea would be to review these with people in your team or your close stakeholders. They are likely to have different views that can stimulate your thinking.

Assign Probabilities

This can be quite tricky, especially if you are not particularly close to the action that could lead to the change. The idea here is to pick a figure based on what you know, think and feel, for each of the main scenarios you believe you need to consider.

As time goes on, your views on these probabilities will change, so each time you review this work, adjust the probability score. Is each scenario more, or less, likely to happen? The reason you need to do this is that you cannot plan action or develop a strategy for every scenario. Making sure to review this will keep you alert to when you need to take new action or adjust your plans. Rather than wait for things to happen, get ahead of the game.

Consider Indicators

If you know what to look for, adjusting the probability score will be easier. The clues could be obvious or obscure. The main point here is that you need to turn on your radar to watch out for things that may mean you have to take decisive action.

To help you on this, take each scenario option in turn and imagine that it has come true. What would you notice that is different? How would people be behaving? What announcements would have been made recently? What other things would be changing? Who would be in favour? What changes would they be making? As you think about these types of questions, start to develop ideas about what things you can observe in the run up to that scenario coming true.

For example, if one of your options is that the Finance Director is going to become the new CEO, several weeks before this is announced it is likely that you'll notice him in increasingly good humour. He'll also start to get even busier, delegating more of his work to subordinates. Meeting with senior people from other parts of the business that he doesn't normally see. Having unusual meetings, perhaps several in one week with key shareholders.

These things could have other causes, so you need to remember that they are only clues. As they are noticed, you may need to do something to investigate a little deeper before you jump to any conclusions and take action. You never know, the FD may be about to retire!

That's all I need to say on the topic of political scenarios at the moment. This is a method of looking from a high level down towards your project or goal. At the end of the chapter, I will ask you to draw conclusions that will affect your strategy development. Before that, I'd like to help you consider the big picture from the opposite end of the spectrum, starting with your goal and looking up. This will help to see things differently and may identify new things that you need to take into account.

Risks and Opportunities

This is a more straightforward approach to improving the quality of your thinking ahead of strategy formulation. You may be familiar with managing risks on projects, and this is exactly the same, except with the addition of considering the opportunities, which few people do actively.

Regardless of the level of skill you exercise, it is always possible that something can go wrong. The severity can range from mildly frustrating to "show stopper." The way people handle these events varies greatly. At the opposite end of the spectrum, other things can happen which can dramatically improve your prospects. Often, at first sight they have only a minor connection with what you are engaged in. Being able to spot these and handle them for maximum advantage or minimum pain is important.

Ironically, those who get caught out here are often the people with the best execution skills. Their drive to get the result narrows their focus, and they can easily lose awareness of what's going on around their project or goal. They easily develop blind spots and miss warning signs because they are so committed (or sold) on the value of their result. So, it is prudent to stop from time to time to look at the risks you may be facing and the opportunities you could make more use of.

The level of detail you need to go into with this is dependent upon the context of your goal and the complexity in and around it. The more complicated your goal, or the bigger its potential impact on the organisation, the more time and effort you need to invest in considering the risks and opportunities. Even in the simplest of goals, I believe that it is worth taking ten minutes to attend to this topic.

There are many different risk management models; however, I like to incorporate the opportunities too. You may be using other approaches as part of your project to manage and control risks and issues. If so, keep doing that and use what follows as more of a personal approach to risk and opportunity management. As with the other approaches in this book, it is very simple to use. Before I introduce it, first assemble a big list of risks and opportunities.

Brainstorm the Risks

Take a few moments to brainstorm all of the things that could go wrong in and around your project. What might happen in the wider organisation which could cause your project problems that make success more difficult? Here are some questions to help you:

- What crises could the organisation face in the foreseeable future?
- What critical resources could be withdrawn?
- Is there something that may change in relevant legislation that might affect what you are doing?
- Are there any significant technology considerations that may impede your progress?
- If your project wasn't to proceed, what would quickly take its place?
- If you were the CEO, for what reasons might you kill your project?
- What could happen that would really damage your progress?

Brainstorm the Opportunities

Now, spend some time considering the opposite; what are all the things that could really help your progress? To stimulate your thinking, consider:

- What are all the big exciting projects going on (or being contemplated) in the organisation at the moment?
- Where are the executives focusing their attention right now?
- If one thing were to happen to make your goal easier immediately, what would it be?
- If I gave you one wish (related to your goal), what would it be?
- What event could happen that would really accelerate your progress?

The Risk/Opportunity Assessment Model

Now you can bring your lists together and prioritise those you need to pay attention to. To do this, review each item on your lists in terms of the likelihood of the event happening, and the impact it will have on your project.

When you have assessed each list, you can then plot the events onto the chart below. This will help you to gauge which ones to explore in more detail before finalising your strategy and plan to achieve your goal. To seasoned project managers, I am sure this may look a little ordinary. What is unusual is that it incorporates all types of events into one place by considering positive and negative impact on your project.

Risk / Opportunity Assessment

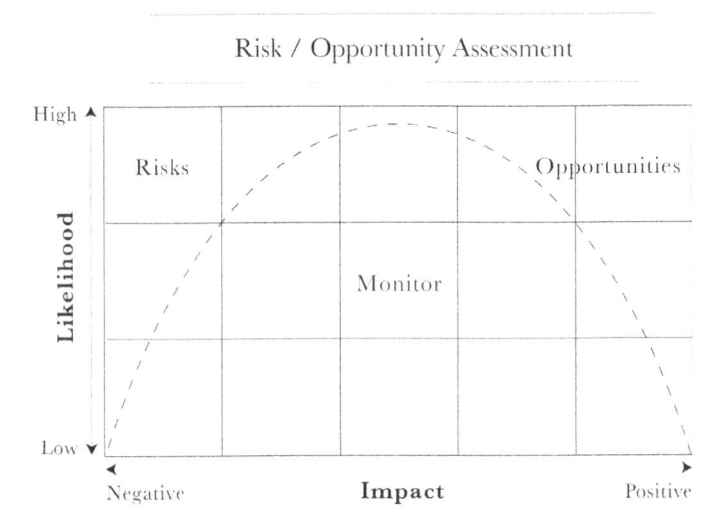

[Chart showing Likelihood (Low to High) on vertical axis and Impact (Negative to Positive) on horizontal axis, with "Risks" in upper left, "Opportunities" in upper right, and "Monitor" in the centre, with a dashed arc connecting them.]

Likelihood

This dimension reflects your opinion on how likely it is that the event will materialise. Although simple in concept, it is more difficult when you come down to trying to assess this probability for events where you have no direct insight or connection. Sometimes, it is necessary at this stage to pause while you get input from other people who have greater knowledge of the situation.

If you are in a big hurry, take a guess and see where it takes you. Then come back later and reassess the position of the risk or opportunity. The information and intelligence you collect in the meantime will help you to become more accurate. If you haven't got it noted down somewhere, there is a significant risk that you'll forget to reassess it.

Impact

At its simplest, this means — how much could each event help or hinder your project or goal? You need to determine the nature of any impact, as well as its severity, and this can become very complex in a short space of time. For significant events, particularly those with high likelihood, this would be an extremely worthwhile endeavour. Unless you really get to grips with what could happen, you could be accused of sticking your head in the sand. Equally, attending to the big opportunities could save you a huge amount of time and energy if they come to pass and you are well positioned to benefit from the event.

Sticking with a practical approach, use your gut feeling to estimate if the impact will be either positive or negative, and how strong that impact will be. As with likelihood, you can seek further input if you need to. Most of the time, however, your first and quick assessment will probably be the most useful and time effective.

Completing Your Assessment

Now it is time to complete your assessment. Make sure and assess each event on your lists (risks and opportunities) in terms of likelihood and impact (positive or negative). Then plot them all on the grid. If you are considering a great many events, use letters to plot each to avoid the grid becoming unintelligible.

Once you have entered each event on the grid, the bigger picture will start to become clear. You should place a high priority on any risks or opportunities outside of the semi-circle. Those with either high positive or negative impact on your goal that also have a high likelihood of happening are worthy of your attention right now.

For each event outside of the semi-circle:

- ° Do you need to take any action right now, either to lessen the likelihood of negative impact or to increase the likelihood of positive impact?
- ° What additional information do you need to get?
- ° Who else do you need to talk to about these?
- ° What implication does this event have on your goal now, or could it have in the future?
- ° Does this introduce new stakeholders that you need to consider?
- ° What does this mean for your strategy to achieve your goal?

Again, I'd like to stress that what I am suggesting here is a light touch risk/opportunity management process. Its primary purpose is to pull you briefly away from the people focus of the Stakeholder Influence Process. Viewing your plans from a risk and opportunity perspective may well make your success inevitable.

One final thought. It would be prudent to reflect back on your political scenarios. It would be nice to see harmony between the two, but in most cases there are likely to be differences caused by the alternative viewpoints you have taken. If there are differences, why might this be? What have you missed?

Key Points

- ° The big picture will inform your decision about which stakeholders to engage with, and which you can safely ignore.
- ° No project or goal exists independently of the host organisation. To ignore that fact is foolhardy.

- It is rarely possible to be precise with this because things are changing so fast. So you have to work on the basis of probability.

- Probability will become more accurate if you remain alert to emerging indicators.

Suggested Actions

- Explore your theories and scenarios with trusted friends and advisers. Their perspectives, opinions and insights can greatly enhance your own.

- Identify several new opportunities to collaborate with others with compatible agendas.

- Summarise your scenarios and risks for future reference.

COFFEE BREAK

Diverging Objectives

The more you delve into the relationships surrounding your project, and uncover what is really going on, the more likely it will be that you will realise that the objectives of your project don't fit. When your project was initiated, clearly the powers that be considered it to be right; however, things change. If you're in this situation:

- Ignoring these changes is dangerous. You may be able to keep your head down and keep working away at what you were asked to do, but will it be a worthwhile effort?

- When you notice these, I'd argue that you are duty bound to do something about it.

- Make use of established change management processes and governance structures to raise the concern.

- Remember to stay focused on the needs and priorities of the host organisation. Projects don't operate in isolation and need to be working for the organisation as it changes.

- Other business decisions may not have factored in the cost implications on what you are doing. Perhaps they should change to accommodate your project costs?
- If you have good relationships with your key stakeholders, it will be much easier to move towards a great decision.

The trouble with this is that many project managers just work on their deliverables and ignore the reality of an evolving brief. This is a very rigid perspective and not part of the way influential project managers should be delivering their service.

CHAPTER 9

Advocates, Critics, Players and Enemies

The bigger picture is essential in helping you to formulate influencing strategy; however, it can also be affected by the position each stakeholder holds on your map. Considering the implications of their positioning is the final step before deciding on your strategy to advance your influencing goal.

This chapter will help you to:

- Understand the implications of different positions on the stakeholder map.

- Decide which stakeholders to focus your engagement and influence on.

- Adjust your approach to influence them more effectively.

If you're in a hurry:

- Skip this chapter. You've probably already got a good sense of how to handle your enemies and advocates, so press on.

- If you're not so sure, at least skim this one.

The variance on the two dimensions of relationship and agreement creates distinct differences to the way stakeholders will behave and react when you seek to influence them. It is important to understand these differences because it can completely alter the way you approach them.

Quite literally, these differences can mean the difference between success and failure. To become excellent at influencing your stakeholders, you need to learn how to adapt and flex to meet their specific needs.

Engaging with Advocates

Advocates are your best friends and wise mentors. So give them the time and respect that position deserves. They will be invaluable in helping you to overcome obstacles and build your understanding of the political reality around your goals and your work.

Although they will probably volunteer to solve problems for you, try to resist this. It can be less effective than you may think, and until you learn to stand on your own feet your career may get stuck. It is on Advocates that you should ideally focus much of your attention.

Remember that you cannot be complacent with them. Sure, they are already in a good position, but are they really advocating your goal? A real Advocate will go out of their way to speak up on your behalf. When you're not at the meeting, they will be proactive in protecting your position and promoting your goal. This is different from a supporter who may stand up for you. Often they only do this when prompted. Advocates don't wait to be asked.

Promoters   Advocates

Advocates

Supporters   Friends

On the Stakeholder Influence Map, imagine dividing the top right box (Advocates) into a 2x2 grid. Advocates actually sit in the top right of this box. Down towards the bottom left you'll find the Supporters. Other distinctions are Friends who really like you, but are not totally convinced of what you are doing, and Promoters who are probably more concerned with goal realisation than in you doing it. They'll promote the result rather than you.

My challenge to you is — how can you move important/powerful people up to top right corner so they become mobilised, proactive and start using their power to make things happen for you?

An obvious first step is introducing them to your vision and taking them through the benefits they will accumulate when you are successful. Consulting with them to find new ways to improve progress on a goal will give you valuable input while also making them more committed to what you are aiming for.

With Friends, the real challenge is getting them more bought into what you are doing and the benefits they will gain. Link your agendas to theirs wherever you can. They may not have thought about it the way you do. The good news is that because of the strength of your relationship with them, they will be open to you and will give you a fair hearing.

For Promoters, they don't need convincing of the merits of your goal — they are already well on-side. What they are not so sure about is you. Perhaps they are not convinced that you are the right person for the job? In which case, focus your time and effort on giving them opportunities to get to know you. A good way of doing this is getting curious about them. Attempt to find out more about their agenda and what they are aiming for. During the conversation, you will have plenty of opportunities to let them see your real talent.

You'll need to work on both the benefits of your goal and your relationship when it comes to Supporters. If you can, take them through the Friends route towards Advocacy. That way they will buy you, and then they will buy what you are doing. There is a slight risk that if you try to take them towards strong advocacy via the Promoter route, they may run off with your goal before they buy you.

Another thing you can do is to step back from the goal you are working at and look at the bigger picture of your relationship. The chances are high that if they are advocating one of your goals, they'll also be advocating most of your goals — and you as an individual. You have trust; the relationship is open and frequent, they like what you are doing, and things will be working well between you two. Does this work the other way too? Are you their advocate? Do you shout about their work, projects and goals?

In my experience, imbalances are okay, but serious differences usually lead to trouble in the relationship further down the road. If they are a strong fan of you for a few years and then they work out that you consider them to be an idiot, it could get rather difficult between you two. Anything you can do that will improve the balance in your relationship will enhance it and make it stronger — for mutual benefit.

Here are some more ideas on how to improve your general position in the mind of your Advocates:

- **Consider competition.** In the mind of an Advocate, who else could they promote instead of you? People are only able to advocate one person or project at any given moment, so who are you competing with in their portfolio of friends? How can you raise yourself on their radar ahead of the others?

- **Become distinctive.** What is it about you that can raise you in the mind of your Advocates? What sets you apart from the crowd? If you can develop your uniqueness, and it is something your Advocate really values, they are much more likely to be shouting about you. So work on your personal brand and personal power.

- **Add more value.** Make sure and continue to add value to your Advocates in the area you want them to promote you in. If you can get yourself into the position of being their resident expert, someone they refer to when they need advice and support, you'll be doing really well. The best way of advancing this is to keep thinking of ways you can enrich your relationship by adding more value.

- **Give feedback.** When someone approaches you as a result of an Advocate promoting you, find a time to thank your Advocate and let them know what happened. Equally, if you learn that they have acted to promote your interests, thank them. Not only will they appreciate the feeling that they have done something good, but it will also reinforce the behaviour and make it more likely they will do it again. It also gives you another opportunity to raise yourself in their mind.

- ○ **Ask them.** Delicately let them know what you want and even negotiate with them to speak out on your behalf. This is usually (and sadly) something that seems to be necessary with many bosses who are missing opportunities to stand up and shout out for their staff. They should be doing this, and if they are not, why not? So, if there is someone who you believe should be advocating you or your project/goal, go find out why not and see if you can motivate them to change.

Engaging with Critics

The great thing about Critics (as I am using the word here) is that you have a good relationship. The only problem is that they don't agree with you. But at least you can talk about it. And this lies at the heart of how to engage with your Critics — leveraging the relationship to negotiate agreement. From this you can see that they are not against you, they just disagree with what you are trying to do.

Part of your preparation to engage could involve thinking about what sort of Critic they are. To help you work out your best approach, here are the main classifications I've come across over the years:

Incidental Critics

These are the helpful people who are generally rooting for you, but on the particular goal you are working on, they've got a problem with it. Believe me, this is great news. Why? Because the strength of your relationship will mean that you can quickly get to the bottom of the problem; find out what they think needs to change and then either change it or negotiate.

You are dealing with your cards on the table and don't have to guess and assume. Provided you ask, you'll almost certainly get accurate insights that you can work on.

Black Hat Critics

Borrowing from Edward de Bono's *Six Thinking Hats*, these are the people who are always looking at what is wrong with everything. They have an insatiable appetite for finding fault even in their closest friends, or especially with their closest friends, because they care so much.

The first step for engaging with a Black Hat Critic is adjusting your attitude towards them. They are okay; it's just they may be a little different in their approach. Genuinely appreciating the value they bring to you and the organisation should then be followed with the realisation that they are doing things that way because they care.

Accidental Critics

These are people who actually agree with your goal in concept, but in the context of the wider organisational agenda they disagree. I'm using the label accidental because the cause is largely outside of your control, or theirs for that matter.

To illustrate, they may think you have a great new product idea that will make the company successful, but they think there are more profitable ways to invest the money you need to develop your product.

Alternatively, they could be agreeing, but now is not the right time because of other issues rolling about the organisation. As with all Critics, it is vital to get to the bottom of why they are opposed to you, then you can start working with facts.

So overall, you need to focus on getting a clear understanding of their objections while continuing to reinforce the relationship. The emphasis you place on these two can be guided by where within the box they sit. The further they are towards the left, the more you will need to focus on building your relationship with them.

Engaging with Players

These are the stakeholders you are never quite sure about. They say one thing (usually they agree with you) and then do another. Their actions don't quite back up their words. What you need to do is dig a little deeper into why they may be doing this, then you can decide on your best approach to engage with them.

Incidentally, Players rank behind Advocates and Critics in terms of general priorities for your time and effort. Enemies usually come last in your task list, but more on that shortly.

So, with one of your Players in mind, take a look at these variants and see which one fits.

Game Player

These are the office politicians. They enjoy the game and seem to think it is okay to lie, deceive and manipulate others in order to get what they want. At the extreme, they delight in playing these games. Game Players tend to focus on small squabbles and deceits. The problem is that these slippery characters are very hard to engage with and need some quite assertive action to put them in their place and keep them away from your work, or at least knowing they can't pull the wool over your eyes.

If you have one of these on your hands, take a look at *Managing the Politics* in Chapter 12 and also refer to the resources listed at the end of the book.

Strategic Player

A more advanced and sophisticated version of the Game Player, Strategic Players have a bigger agenda they are working towards. They are carefully positioning themselves and their projects. If they are transparent about how your goal threatens their agenda, it could work against their interests.

Instead of being open, they play their cards close to their chest, waiting for the right moment to play their trump card. Be careful with these, they have a plan, and they probably have a stakeholder map too. Pay special attention to the earlier chapters on political agendas and risk management. Work with your Advocates (and even your Enemies) to gain more political insight into what their plan might be — then decide how best to tackle them.

A great question to ponder is what does their stakeholder map look like? A head-on, forthright approach, is unlikely to work with them because of their skill in martial arts — you'll be on your back before you work out what's happening. So match up to their sophistication and then work on the relationship.

The purpose of this book is to turn you into a strategic *influencer*, not a strategic *player*. I'm at pains to encourage you to develop strategies and plans to achieve your goals. This does not necessarily make you a Player to others. If you also buy-in and implement my encouragement for fair dealing, building high-trust relationships and being open and honest, others will know exactly where they and their projects, stand in your mind.

Puppet Player

These are generally well-meaning people who are genuinely in agreement, but are permitting other people to pull their strings. It could be that they have other people exerting powerful influence on them, so they are sincere in their agreement, but for political reasons they are unable to deliver on that agreement. Their masters can control their actions.

Trouble is, the poor quality of your relationship means that the truth is likely to be well hidden. With these, it is best to try to focus on building friendship within your relationship so they can gradually open up a little more.

Submissive Player

Not everyone likes or can handle conflict. Your impressive ability to assert your views can intimidate people if you are not careful, and some may find your approach aggressive and/or insensitive. So rather than face up to you with their disagreement, they avoid raising the issue because it appears to be the more palatable option. You know that it is important to face up to difficult issues, but they don't think quite like you.

If your stakeholder is this type of Player, go easy on them. Tone down your assertion and draw them out. Make them feel comfortable and, whatever you do, don't immediately and loudly challenge their disagreement once they do build up the courage to say what's on their mind. You could well be missing out on some valuable insights, so gently encourage them out of their shell.

With all types of Player, your focus should be on building the relationship (many more ideas on this in Chapters 11 and 12). Whatever the reason, if they are not being honest and open with you — you'll be left to guess, and I think that you're probably the sort of person who prefers to be straightforward and deal with facts.

Engaging with Enemies

In an earlier chapter, I mentioned that it is often not a good use of your time and energy to engage with Enemies. It does not mean that you should do nothing, but you probably need to err on the side of indirect action rather than engagement. The reason for this is the combination of a poor relationship and open disagreement. When you've put someone in this box, remember that it is usually a provisional assessment and that the title Enemy is used to provoke your thinking, not label them as horrible people.

That said, with anyone here, you clearly have some big concerns, and if they are powerful people, you will need to do something, and you will need to judge this one well. The general approach forms a mini process which offers no guarantees, but will at least give you a structure to determine what course of action is best for you.

1. Recognise the detail of the problem and become determined to do something rather than avoid or dodge the issue (most of the other chapters in this book can help you here, but particularly the ones about agendas, power and risk management).

2. Seek wise counsel from your Advocates and friends to get their views and ideas about the problem. You can be open with them and get truthful input.

They may also be able to help directly in ways that you cannot.

3. Carefully probe around your Enemy to test your theory of what is wrong. Then pull back and give yourself time to consider your options.

4. Explore ideas that could reduce the negative impact they could have on your goal or the likelihood that they will successfully hinder your progress.

5. Are there any opportunities available for you to build a better relationship with them? Many Enemies are actually people who don't know you very well or understand what you are trying to achieve. Letting them get to know you at a personal level can make a remarkable difference, so it is worth a go. And that works both ways too — perhaps you just don't know them well enough.

6. What can you change about your proposal that could shift the basis of their disagreement? If you can remove rational disagreement, that just leaves irrational disagreement — and nobody likes to appear irrational.

7. Having considered these ideas, what's your plan? Make sure you have a Plan B too and be ready to change course if things start to get worse. Test out your thinking with your Advocates.

8. Keep your Advocates in the picture so they are ready to help spontaneously when they can. If they know you are about to go and ruffle a few feathers, they can be ready to stop a counter-attack at the next board meeting. If they don't know what actions you are taking, they may be unprepared and miss an opportunity to stop it.

As you do this, you need to remain realistic about your hopes for engaging with Enemies. They always present challenges, and remember that if you are able to counter their negative impact with the help of your Advocates, that is often not only the most likely strategy for success, but also the least stressful.

You would also be well advised to be graceful and kind when you succeed. Beating an Enemy can make them more determined next time — so tone down your victory parade.

Proceed with caution, but smile; you're at the cutting edge of organisational performance. If you can turn an Enemy into an Advocate, your level of skill will be something to be proud of. If you can do this without bloodshed — awesome!

In closing, please remember that people move between the boxes for a number of reasons. They could shift because you made a mistake in your assessment, your engagement campaign works, something changes elsewhere in the organisation, or maybe they just reassess your goal and think it is worth backing. You cannot assume that these changes will always be positive, but you must stay on top of all these changes as you go. That is why the Stakeholder Influence Process includes regular reviews.

Key Points

- Focus plenty of attention on your Advocates. Don't take them for granted and make full use of their support.

- Moving people within the Advocates box can be fairly easy.

- Critics are there to challenge you and make you work hard to refine your arguments and proposals. Welcome this as their contribution to your success.

- Make it a priority to build relationships with Players so that they will open up and tell you the real reason for their resistance.

- You need to understand the threat posed by Enemies, but you don't necessarily need to engage with them.

Suggested Actions

- Move on to the next chapter to finalise your strategy and plan. I know you're itching to engage with your stakeholders now, but just pull your plan together first, especially if success is really important to you.

COFFEE BREAK

Project Governance

Policies and processes are implemented around (or above) a project in order to configure the power to maximise the chances of the project realising its goal and benefit. They are usually driven by the organisation seeking to control and monitor what is going on; however, the external environment, particularly legislation, is creeping into this space rapidly. Regardless of the cause, project managers need to become accustomed to the presence of governance and work with it rather than push against it. Here are some questions to ponder:

- To what extent are you clear about the exact nature of the governance approach that has been put in place?
- Why has governance been put in place for your project?
- Do you know who the most influential people are in these processes?
- Are there any political motivations behind anything that is happening?

- What role do you play, or should you play, in the governance processes?
- How powerful is the governance in real life?
- To what extent does it help or hinder you?
- How can you make more effective use of this source of influence?

One thing for sure, you won't be able to avoid it. Trying to dodge it is likely to be short-lived, and the more evasive you have been, the harder it will fall. It is there to stop you doing your job; however, it could seriously steal your time with the amount of bureaucracy, much of which is far more discretionary that you may think.

CHAPTER 10

Step 4: Plan

Assuming you have been diligent in your application of the preceding steps, now is the time to pull it all together and make quick decisions about the actions you need to take in order to maximise progress towards your chosen influencing goal.

This chapter will help you to:

- Pull together the threads of your thinking into a coherent strategy.
- Focus on the most important things you need to accomplish in order to reach your goal.
- Simplify things so you can maintain your focus.
- Get ready to engage your stakeholders.

If you're in a hurry:

- And you have a clear notion of the key things you need to achieve to progress towards your goal, just write them down and work out a plan of how you are going to deliver each.

- If you're not so sure what to do for the best, you'd best read on.

What is a Strategy?

The word "strategy" seems to be used everywhere in business today. Put simply, what I mean when I use this word is the main steps you are going to take over a period of time in order to achieve your goal.

It could be useful to think of this in terms of the stepping-stones that you need to move safely across from where you are today to where you want to be tomorrow. However, there is a slight problem with that analogy. It presupposes a linear path with one step leading to the next. When it comes to influencing, you often need to move on many different paths at the same time, with a view that they each contribute towards the end goal. The great thing about this is that should one of your paths slow down, it doesn't necessarily prevent you from reaching your goal.

Instead, think of your strategy as a number of individual goals that will move you nearer to your goal, and perhaps contribute to each other too. Each of these goals may be of sufficient size and complexity to warrant applying the full Stakeholder Influence Process in their own right.

So, the output from this step in the process could be called your "campaign" of influence. Using the word campaign helps to position this step in your mind in much the same way that a political candidate may develop their strategy to become an elected representative. This is a helpful mental image because it will encourage you to stay focused on the main task in hand — influencing people.

In the majority of cases, it is of critical importance to your influencing goal that you win hearts and minds. Of course, you also need to get your facts straight and plans agreed, but usually, when using the Stakeholder Influence Process, the main job is getting people to buy into your ideas, concepts and vision.

General Approach to Strategy Development

A good strategy to influence a goal needs to be able to address the following key questions:

1. How will you build more support for your goal?

2. What do you need to do in order to convert or neutralise opposition to your goal?

3. How can you manage down the risks and attach yourself to opportunities for progress?

4. What can you do to make it happen faster?

5. As you make progress, how are you going to protect it?

6. Is your strategy simple and memorable?

These are all pretty straightforward to understand and can be used to sense check the strategy that is emerging. You can read about these in more detail in my last book, *Influential Leadership: A Leader's Guide to Getting Things Done*. The last point is worth expanding on a little here.

Having spent many years being responsible for international strategy earlier in my career, one of the big obstacles to successful implementation is that people forget what the strategy is. With everything else that is going on in a busy working day, all the problems and issues that people are facing, little wonder that strategy gets forgotten.

In that role, I went to great lengths to keep it current and make sure everyone kept the strategy at the forefront of their minds during the working day. If the strategy is remembered, it will help each person to make small decisions during the day that continue to move things in the right direction. For example:

- When an email comes in asking you to do something, what impact does that have on your strategy?
- When someone phones up, how can you use that contact point to push forward your strategy?
- At any point during the day, are you moving towards your strategic goals?
- Could you be moving faster towards them?
- Which meetings are really necessary to attend?

Those who make the fastest progress towards their goals are usually the ones who remain focused on their goals. If you are like most people I meet, you probably have far too many goals to focus on, so how can you possibly progress on all of them?

The simple answer is that you can't — something will fall by the wayside. Because of this, you have to work hard at simplifying everything you are doing so that you can keep the critical things in mind day to day.

Therefore, on the goal you are working on at the moment, keep it as simple as possible. Make sure the steps are easy to understand and worded efficiently. Then keep a prominent reminder of what they are and return to them every day. Do this for every part of your work, even if that means putting some things on the back-burner for a month or two.

The Threads of Your Intelligence

Keeping a particular goal in mind, reflect back on the topics already covered in this book by answering these questions:

- How will the achievement of your goal affect the power structure of the organisation?
- What are the political purposes of your goal or project?
- Who are the key stakeholders and what are their personal agendas?
- What is the bigger picture into which your goal or project must fit?
- To what extent does each stakeholder agree or disagree with what you are trying to do?
- Who are the biggest winners and losers?
- What's the quality of your relationship like with each stakeholder?
- Who needs to move on your stakeholder map to maximise your progress?
- What are you going to do now?

You may not yet be in a position to answer the last two questions, so here are a few more pointers to stimulate your thinking.

Moving Stakeholder Positions

At a high level, to accomplish your goal, you need to be able to move sufficient power into the Advocates box. Once this box has sufficient power within it, you should be able to start relaxing — even if there are still powerful forces against you, your friends should be able to help you win through. In real life, it is usually difficult to achieve this, but it does form a basic principle — you need to be moving powerful people to the right and upwards. Problems and issues usually lie in the opposite corner around your Enemies.

In essence, this means you need to be continually looking to build good working relationships with your stakeholders, while at the same time increasing the level of agreement that exists for what you want to achieve. The next two chapters will go into more detail about how to actually do this. Here you need to be setting your sights on who you need to engage with. Worry about how to do this a littler later.

As you think about the bigger picture, look for ideas about the major things you need to achieve to get there. Below are a number of points to stimulate your thinking. Some of them reiterate what has been covered in previous chapters because they are vital to consider right now.

○ **Concentrate on impact:** Focus on those who can have the greatest influence on what you want to make happen — the people who can help or hinder in a big way. There is no point spending time and energy on the minor players unless they are your only route through to the real power brokers.

○ **Shift the greys:** Stakeholders that ended up in between the main boxes (perhaps because you don't know them very well) need to be moved fast, especially if they are very powerful. It is much better to be sure that you have a powerful Critic than to be lost in the dark. Once they are in the box, then you can work out how to engage with them.

- **Move up and right:** As mentioned earlier, the general principle is moving people vertically up the agreement dimension and horizontally right along the relationship dimension.

- **Movements inside the boxes:** Consider opportunities to move people within their box. As discussed in the previous chapter, moving a powerful Critic from the very bottom of their box to the top half may be sufficient for your purposes. By way of illustration, you might imagine that a Critic could have ten reasons why they believe you should not succeed with your proposal. If you can win them over on eight out of the ten, they might be happy to just let you pass through. Similarly, an Advocate, who is in the bottom left of their box, could be quite easy to move upwards and right. Remember that the higher somebody moves up on the map, the more action they will be prepared to take to help you. An Advocate in the bottom left is often an opportunity missed.

- **Advocates are top priority:** This is often overlooked when considering influencing strategies, because they are already on side and thus don't need to be influenced. In addition to finding ways to get them to take more action, you can also work to direct their action towards specific individuals in the other boxes. They will have different relationships with these individuals, and they may be able to exert more power to remove problems that you are facing. The other great thing about Advocates is that, because you have an excellent relationship with them, they will be able to offer you honest and practical advice to help you overcome the challenges that you face. Make full use of these powerful friends.

- **Critics make great opponents:** Because you have a good relationship with people in the Critics box, this means you can potentially negotiate with them. It also means that you both have a transparent way of transacting business — in fact, you probably both know exactly where you stand on the goal in question. I often refer to Critics as "best friends," because they will help make your proposals and ideas more robust and successful. There is nothing wrong with having opposition as it will push you to be your best.

- **Ignore your Enemies:** You need to be careful with this one, and I realise that I am arguing against the famous maxim "Keep your friends close and your enemies closer." In my experience of working with people using the Stakeholder Influence Process, Enemies are often the first place for them to focus — but also the most costly in terms of time, effort, energy and stress. The problem is that since you have recognised the poor quality of your relationship with them, and if you are correct in your assessment, your potential to shift the relationship to the other side of the map is limited. I am not saying don't try, instead be careful you don't place too much emphasis on these troublesome characters. Far better would be to favour your Advocates and enlist their support in minimising the risk/damage that Enemies could cause. It is sometimes amusing to observe that Enemies often voluntarily move on the map if they feel they are being ignored.

- **Timing and sequence.** Another thing to bear in mind is that you cannot fire off in all directions at once, nor should you. As you consider what people need to move, think about the optimum sequence of your activity. Choosing one stakeholder to engage with first may make the others easier to engage with. There may also be an optimum time to take action, for instance waiting until other things have happened in the organisation before making your move.

- ○ **Remember the indirect routes:** Sometimes the most effective influence is created through other people. Often, the most important people to get agreement from can only be accessed indirectly. In which case, look for others who can do this for you. Enlist their help and support and make them Advocates.

Of course, you may need to consult with others when you are working out what direction to take, but don't take too long about it — you can always adjust your trajectory later as you learn more. To reiterate, the most critical part of the process is getting moving and taking action.

Exercise

Pause for a moment and note down the key ideas you have for what you need to make happen in order to achieve your goal. What paths do you need to move forward and what sub-goals do you need to achieve?

Here is a couple of simple examples taken from my clients.

Sue wanted to bring in a new stock ordering process. It was expected to yield big benefits, but she was struggling to get it moving forward positively. When she mapped out her stakeholders and considered what her strategy should be, she realised that many of the Players were also Area Sales Managers. Given their power, she realised she needed to focus her effort on winning them over. If she could do that, everyone else would fall into line. Her headline campaign strategy became:

1. Get invited to the next ASM meeting.

2. Build a communication pack focusing on how the process would lead to greater sales.

3. Influence the IT people to change the process in order to make it more sales friendly.

4. Become a champion for the sales teams.

Even simpler:

1. Go to ASM meeting.

2. Produce Sales Pack.

3. IT make friendly Sales

4. Champion Sales

Leroy's goal was to get the business leaders to really buy into the benefit his change team could bring to the organisation. He realised that some were Advocates, while most were opposing, mainly on account that they viewed his work as excessively bureaucratic and that it got in the way of the real work. He noticed that one of the Critics (Bill) was fairly new to the business and was less powerful than the other leaders, but very ambitious. Leroy's strategy became:

1. Get Bill to recognise how working more closely with the change team could not only help his business, but also help him to become successful and powerful within the organisation.

2. Then, advocate Bill at every opportunity to senior management, pointing out his successes linked with adopting change management approaches.

3. *Influence Peter, the MD, to become more active as an Advocate and to highlight the causes of Bill's success (process improvement) to the other leaders on his team.*

The simple version:

1. *Make Bill powerful.*

2. *Promote Bill's achievements.*

3. *Get Peter to promote Bill's success.*

Now, what's your headline strategy to achieve your goal?

Building your Plan

I'm going to assume that you've got your headlines now. If you haven't, you're going to need to talk to others around you pretty quick and get more help that way.

Once you've got your headlines, simply write down all the actions you and others need to take to achieve it. Go into the level of detail that is appropriate for the step you need to take. If it is more than a few items, you'll probably need to be writing it down and making lists of what to do and when.

Another assumption I'm going to make is that you can do this quite easily because I'm going to move on now to how to engage with your stakeholders to build stronger relationships and greater agreement.

Key Points

- Aim to move sufficient power into the top right of your stakeholder map so that your success becomes almost inevitable.

- You don't have to please all of your stakeholders.
- Keep your strategy simple and memorable.
- Work hard to identify the main steps or achievements that will contribute towards your overall success.

Suggested Actions

- Get your line manager, coach or mentor to challenge your strategy and plan. In fact, why not get them all to contribute.
- Keep your team involved in developing your strategy if appropriate. This is really important if they are going to be involved in the execution of your plan.

COFFEE BREAK

Steering Committees

Love 'em or hate 'em, steering committees are a key part of most sizeable projects. Should you be worried, or are they there to serve? While you're having that quick coffee break, think through your answers to these questions:

- What is the purpose of your steering committee?
- Who are the most powerful members on the committee?
- What is each individual's remit or role?
- How does your attitude towards the committee affect things?
- What links do key members have with other projects?
- Are you reporting to them in the most effective manner?
- Do you have a high-quality relationship with each member?

◦ What opportunities are there for making more use of their influence?

Many project managers report to their steering committee rather than the committee reporting to them. While it may seem a little odd, why isn't it the other way around? Once upon a time that is exactly how it operated, the steering committee providing support to the project manager. Could you subtly or even overtly influence the terms of reference of the committee to make them more effective, for the good of the project?

CHAPTER 11

Building High Quality Relationships

To create high levels of influence, you need to have high-quality relationships. It doesn't matter how clever your strategy and plan, without relationships that are based on trust, openness and a spirit of collaboration, you will continue to struggle. Once you get the quality right, you'll find gaining influence far easier, and you'll also enjoy your work so much more.

This chapter will help you to:

º Analyse what is working, and not working, in your existing relationships.

º Appreciate the critical factors you need to establish in new relationships.

º Build trust more effectively and accelerate towards the most productive relationship level.

º Reflect honestly on your own contribution to poor relationships.

- Learn what action needs to be taken to improve things quickly, even in the most complex relationships.

If you're in a hurry:

- If you are facing difficulties in major relationships, you really should read this chapter. It will probably show you why and what you can do to fix things. Then it will save you a great deal more time.

This relationship aspect of the Stakeholder Influence Process sets it apart from most other approaches to stakeholder management. If you accept that influence is a very human (psychological) and social (human to human) phenomenon, relationships will have to figure highly in your priorities.

When you were reading about the stakeholder influence map, I asked you to assess the level of trust, openness and frequency of contact with each stakeholder. That is sufficient for the mapping exercise, but if you need to move stakeholders along the relationship dimension, a deeper look will be necessary. This is particularly important if you need to improve the position of Players and Enemies.

Over the last few years, I have been working with organisations to help them develop their relationships with suppliers and strategic partners. As part of this work, we researched the characteristics of highly successful (and not so successful) alliances and partnerships. We noticed three key themes that were symptomatic of excellent relationships. Happily, these themes easily translate into individual relationships too, so this work is a useful way to learn how to improve any stakeholder relationship.

The great news is that they are pretty easy to understand and take action on — you don't need to go into mediation to make big strides in improving the quality.

- **Trust and Credibility:** The confidence that you can rely on each other to match delivery with expectations at both a personal and professional level.

- **Communication and Influence:** The confidence that everyone can clearly state their views, opinions and ideas with an equal opportunity to influence each other.

- **Problem Solving and Conflict Resolution:** The willingness and ability to face the difficult issues and work to move things forward in a proactive and constructive manner.

If you have an abundance of these in your individual stakeholder relationships, you're also likely to have openness and frequent contact. The sections that follow will go into more detail about the evidence that will indicate the state of each of these themes in your relationships.

Trust and Credibility

These two words go hand in hand. If you think, someone is credible (for whatever reason) you are likely to trust them. Credibility is the culmination of a wide range of factors that build the overall impression in the mind of the observer, including qualifications, competence, reputation and performance. When experienced, it is self-reinforcing and ultimately enhances the predictability of the individuals concerned — it builds trust.

Because credibility is so important in relationships and to influence, it was the first source of power discussed back in Chapter 4.

Positive Indicators

When there are high levels of trust and credibility in a relationship, each party is likely to:

- Tell the truth and the whole truth (are open with each other).
- Share sensitive information (even if this could put them at a disadvantage).
- Have no secrets (obviously within commercial boundaries).
- Be predictable and reliable, or even dependable.
- Do what they say they're going to do.
- Give bad news early and avoid "surprises."

Negative indicators

On the other hand, where trust and credibility is lacking, you would notice people:

- Being hesitant, cautious and a little suspicious of the other person.
- Lying and withholding information when it is to their personal advantage.
- Misleading others about their real agenda.
- Creating false deadlines, or moving them for their own convenience.
- Asking different people the same question until they get the answer they want to hear.

- Criticising others behind their backs.
- Providing inaccurate, misleading or false feedback.

Communication and Influence

It is self-evident that communication is an important element of relationships, but the presence of influence in this theme may be unexpected. Usually, you will focus on influencing others and perhaps not wish others to influence you. The idea that it is useful to you if others can influence you may require a little explanation.

There are two key reasons why I believe this is a good thing. Firstly, if you want to have a mutually beneficial relationship with someone, you will have to accept that they may need to influence you at times. To not accept this is perhaps a little arrogant — can you always be right? If they have a good idea that could help you both to become more successful, it would be in your interests to hear it, even if it's contrary to what you currently believe to be the right course of action. If they are convinced that it would benefit you to take a different course of action, it would be madness to deny them the chance to influence you. Fostering two-way influence maximises the potential benefits for all concerned.

The second reason is that if either side in a relationship does not feel they can influence the other, they are likely to feel helpless or powerless. One side of the relationship is dominating the other. If you have the upper hand, of course, this will work for you — at least initially. But, as time progresses, the relationship will deteriorate and quickly drag down the levels of trust. You may have experience of what it is like to be in a relationship where you are unable to influence the other side. It doesn't feel good does it?

Positive Indicators

When there are good levels of communication and influence, you can expect people to:

- Share views and opinions.
- Take time to listen to the other side's views.
- Share a broadly equal sense of power.
- Negotiate fairly in a way that appears more like problem solving.
- Clearly understand the other person's position, concerns and agenda.
- Proactively give direct feedback — straight talking.

Negative indicators

You'll notice people:

- Not attempting to influence the other side.
- Misunderstanding requirements, requests and deadlines.
- Demanding compliance from the other side.
- Complying with demands without challenge.
- Dominating the conversation, or saying little.
- Escalating issues rather than dealing with them.
- Showing high levels of stress.

Problem Solving and Conflict Resolution

Once pointed out, this theme is an obvious requirement for a successful relationship. I cannot think of any relationship that at some point has not had a few problems that included at least a bit of conflict. To be called a relationship, there must be at least two people involved. Each individual will be seeking to influence the other person, so it is inevitable that at times there will be differences of opinion. If these are not handled well (with good Communication and Influence), a dispute is likely to ensue.

So, being able to deal quickly and openly with these disputes becomes a critical element of a successful relationship. I am wise to the argument that if you have a relationship with perfection on the first two themes, you shouldn't need to use problem-solving and conflict resolution. But, even in perfect relationships (and I am sure some get close), it is always handy to have these capabilities available so you can deal with the worst-case scenario, should it occur.

Positive Indicators

People will:

- ° Proactively take the initiative and raise issues constructively.
- ° Follow a clear process to remedy any issues.
- ° Accept responsibility for any failings from their own side.
- ° Want both sides to make the right decision — together.
- ° Give adequate time for discussion and to resolve any issues.
- ° Acknowledge and respect the needs and rights of the other person.
- ° Look for win-win solutions.

- Seek to build constructive ways forward.

Negative indicators

You'll notice people:

- Not facing up to difficult issues.
- Not returning calls/emails from the other party.
- Taking criticism too personally and getting defensive.
- Having a "tit for tat" attitude.
- Behaving in a belligerent, stubborn or childish manner.
- Avoiding responsibility for their own contribution to the problem.
- Hoping the problem will resolve itself.

Assessing Your Relationships

It is a fairly easy exercise to reflect on your relationships and work out roughly where they are by using the indicators above. My caution to you is that this will be your perception, and other people in the relationship may have an entirely different view. When working with clients, especially on big-ticket relationships, almost invariably one side thinks the relationship is much better than the other side does.

Therefore, when you are doing this, keep an open mind and perhaps, go and talk to the other person or organisation involved.

Think of a stakeholder you'd like to improve the relationship with.

Consider each theme and work through the indicators.

- To what extent do you notice these in your stakeholder?
- Be honest, how often do you do these things?
- What might be causing these behaviours?
- On balance, how would you rate the theme on a scale of 0-10? Consider 10 to be exceptional.
- How does this help or hinder your relationship?
- What score would you like to achieve on this theme?

Test your thinking with colleagues or even your stakeholder.

Make a commitment to improving the scores.

The exercise above can be very illuminating in a very short space of time. It can give you more objective insights into how the relationship is working, and it can focus your attention on the things you need to do to improve the quality. It will also help you to improve the quality of other relationships.

If you are working on relationships with many people involved, such as suppliers, alliances and so on, you might like to consider the Collaboration Survey that we developed at The Gautrey Group. This is a simple online survey based on these themes and can be implemented very quickly in large relationships. Find out more in the resources section at the end of this book.

Now, assuming that you have work to do in your relationships, here are some ideas on how to improve the quality within each theme.

Strengthening Trust and Credibility

Since I've already talked about credibility in the chapter on power, I'd like to focus on trust in this section. The key to strengthening trust is to understand what it is, how it works and what you (and others) do to maximise trust, or bring it crashing down.

Trust is a complex concept that can easily be simplified into the extent to which you can rely upon someone (or something). Will they do what you expect them to do? Can you rely on their word? Can you predict what they will do in a given set of circumstances?

There are also various levels of trust. You may be able to rely upon someone to tell the truth; however, you may not trust them with your life — particularly if their own is also in danger. Similarly, you may trust them not to share sensitive information about you around their network, but will they be able to resist a really tasty piece of gossip?

When you meet someone for the first time, you are both carrying a set of assumptions, experiences and beliefs about trust. Some people "trust until proven otherwise," while others "distrust until proven otherwise." Additionally, the person you are meeting will probably have some intelligence about you — what they've heard from friends and their network. Or, they may ascribe opinions to you based on stereotypes.

At the start of the relationship, you will both be assessing each other. To what extent can you trust them? They will be doing the same. Healthy development of trust is a progressive test — share something sensitive (but safe) and see what happens. If that feels okay, share a bit more. Each time it works, a higher level of sharing will take place, confidence grows in the relationship, and the benefits can start to grow rapidly.

Problems arise when the level of trust is badly out of balance. If one person shares excessive amounts of personal information too soon, the other will become wary because of a fear that they cannot keep anything secret — they are too trusting! Equally, problems may emerge if one party is starting to share sensitive information, but the other one isn't. They will soon begin to wonder, "Why don't they trust me?"

Once an initial working level of trust is established, building more trust involves progressively sharing more. This enriches the relationship and helps both parties to gain increasing benefits. However, it is vital to keep working at it and avoid the risk of mistakes causing a problem. A maxim worth remembering is that "trust takes a lifetime to build and a moment to lose."

People who are considered to be *trustable* tend to demonstrate these attitudes and behaviours:

- Have an open mind to the views and opinions of others.
- Show genuine concern for other people.
- Are open about their own position, even if others may not like it.
- Encourage openness and honesty.
- Congratulate and reinforce straight-talking.
- Do what they say they are going to do.
- Manage expectations if they realise they cannot meet a commitment.

People who come across as *untrustworthy* will demonstrate the opposite of all of these. Few people are at either of the extremes, but knowing where you may sit could be very useful as preparation for deciding what you need to do differently.

Before you move on to the next section, explore your experience of trust by thinking of someone who you trust and answer these questions:

- Why do you trust them?
- What do they do that inspires trust?
- What do you trust them with?
- What don't you trust them with?
- Do they trust you? Why?

And what about someone you don't trust? Reverse the questions above and think about this too.

Once you've understood how trust works and how it contributes to credibility, you are likely to be able to quickly develop ideas on how you can strengthen it in any relationships — just look back at the indicators. Here are some more ideas which you may find useful:

- Find safe ways to be more trusting.
- Discuss the subject of trust and see what it means to them.
- Clarify what demonstrates you are credible.
- Don't tell them things you shouldn't, and tell them why if they're asking.
- Don't dodge apparent lack of trust, on either side. Get it out on the table and explore.
- Depersonalise trust, contextualise it instead.
- Never say, "trust me."

- When you spot deceit, open the debate constructively.
- Replace trust with the word confidence. "What I need to be confident is ..."
- Take commitments seriously.
- Manage expectations. If you can't deliver on your commitments, discuss it.
- If you can't tell them something, tell them why.

Strengthening Communication and Influence

The first thing to appreciate is that by far the best way to improve on this theme is to make sure you have lots of Trust and Credibility between both sides of the relationship. The first theme forms an unavoidable prerequisite to successful Communication and Influence. Think about it. If you don't trust someone, the more they communicate, the harder they try to convince you, the more suspicious you will become. Similarly, it is very difficult to influence someone if they don't trust you because they will be suspicious of your motives.

To be blunt, if you haven't got the first theme covered, you may well be wasting your time trying to strengthen Communication and Influence.

Communication is a topic that has received extensive coverage in a host of other books, so I'll keep this part succinct. If you want more insight into the communication side of relationships, I've listed a few of my favourite books in the resources section at the back of this book.

- Communication is a message you want someone to receive. People tend to communicate in their own words and not in the words their stakeholder will readily identify with. Speak their language, not yours.

- How will you know they have received your message accurately? Always find ways to check understanding so you can adjust and refine your messages. This establishes a two-way process and can also offer them the opportunity to send their own messages to you.

- Communication is never perfect. People always seem to clamour for more, yet don't play their part in consuming the well-intentioned stream of messages. Then they complain of too much communication. Finding the right balance seems to be more art than science, but you can tip the scales in the right direction by asking people specifically what they want, when they want it, and how they want it.

- Consistent approaches/processes improve consumption because people learn what to expect and where to get it. So, if you are producing regular updates about your project, stick to the same format, design and structure.

- A picture paints a thousand words. More people than you may realise, think in pictures.

- You want to communicate to your stakeholder. Perhaps they want to communicate with you too? Think about being a role model and helping them to tailor their approach to match your needs, then turn the tables and find out from them how they would like to consume your messages. This makes effective communication an agenda item in the relationship.

Turning to the subject of influence in this theme, a fundamental point to stress here is that you need to bring the level of influence into a more even balance between the two sides in the relationship. Never equal — but to a level where each feels they have the ability to get a fair hearing if there is something they feel strongly about, and that the other party will listen, understand and bring this into account when making their final decision. This lies at the heart of healthy collaboration.

One of the big inhibitors of achieving this balance is the subconscious (or at least unspoken) perceptions of power that each side has of the other. As I explained in Chapter 4, power provides a potent shortcut to influence. One of the side effects of developing great power is that others yield quickly, sometimes too quickly, and without making an attempt to counter-influence because they think it is pointless.

Since the idea here is to build stronger relationships, the relative power of each side, both actual and perceived, can be explored to look for the inequality. This will help you to identify things you can do to adjust the balance (it might be worth taking another look at Chapter 4 before you move on here).

Try asking yourself (and others if appropriate) who is the most powerful — you or your stakeholder? See if you can also determine why that is the case. Remember to consider a wide range of different sources that build up each individual's power. Also, factor in the context in which you are both working. In some situations, very powerful sources are virtually useless. The opposite can also be true. The only power which works consistently regardless of context is the type that comes out of the socket in the wall.

The next question builds further. Does your stakeholder perceive this differently from you? It might be helpful to look for evidence of the way they react to your influence attempts and how they attempt to influence you. If they are letting you walk all over them, they clearly believe you are more powerful, even if you don't see it that way.

The type of influencing tactics they use may also give you valuable insight. If they are using Ingratiation, Personal Appeals and/or Pressure, they could be feeling a little powerless right now. The more they use Consultation and Inspiration, the stronger they'll be feeling (I'll go into these in more detail in the next chapter).

Large imbalances of power and influence between parties to a relationship need to be dealt with swiftly before it causes major problems. Having said that, I'd wager that if the inequality is big, you've already got problems on your hands.

If you think you are more powerful than your stakeholder, consider these ideas to help them increase their influence with you:

- Share the insights in this chapter with them. Learn together the importance of great relationships. Even at a cursory level, it will help them to understand your motives and approach and build their confidence.

- Encourage them to share their opinions. If you don't like what they say, don't attack or become defensive. You have only one chance to make this idea work, because next time they'll have learned not to believe you and will keep their head down.

- Tone down your references to powerful friends, veiled threats, oodles of charm and charisma.

They've already got the message, now is the time to connect with them and stop showing off.

- Check out how you come across. High levels of drive and determination are interpreted differently by others. You may think that it is a normal and straightforward way of operating; others will just find it intimidating. Communication is about tone, expression and a thousand other unspoken signals.

- Boost your regard (demonstrably) for their power sources. Praise them and help them to feel stronger than they are currently feeling. Don't pretend or embellish beyond reason or they will start to wonder what your game is. Appropriately done, this can work wonders for their confidence.

- Let them know when they have successfully influenced you. It is all too easy for highly influential people to miss this opportunity to give this aspect of their relationships a little lift.

- If it is relevant due to the complexity of the goal you are working towards, consider instituting processes or procedures which facilitate fair involvement for all the parties, particularly when it comes to problem solving, but more on that in a moment.

- Think very carefully before you reject their influence attempts. It may be the right decision to take, but remember that it is all too easy for the other side to see rejection and failure. So bear this in mind when you communicate your decisions. If you can help them understand your process, the reasons why, and also thank them for playing a valuable role in pressure testing and challenging, you are much more likely to build a stronger relationship from your powerful position.

And if you're the one who needs to become more influential:

- Do a reality check. When things are not going your way, it is natural to look for something or someone to blame. The actual level of influence you have in the relationship might be good, but for the appropriate reasons it's not working at the moment. You cannot be right all the time — can you? So, check in with a few wise friends to make sure you are not imagining your lack of power and influence.

- Look more deeply at the power dynamic. Often, powerful people bring along the power that works for them elsewhere. This power may not be so useful in the current situation. Likewise, sources of power that you have that don't normally give you an edge could be more useful here. If you spot opportunities, start to figure out ways to bring attention to the sources that should hold sway.

- Decide the level of influence you must have to make the relationship right for you. How much would you have if you could? How does this compares with the current position? Put another way, what have you got, what do you need, and what would you like to have? If you can come up with specific examples of recent events to illustrate this, it will help you prepare for the next idea.

- Find a way to raise the subject with your stakeholder. Get it out in the open. They may not realise it's a problem, nor that it could be working against their interests. This is much easier to say than do. Careful consideration of the other topics in this book should help. For instance, the work you have done on understanding their agenda. In essence, this idea is about influencing them to allow you to have more potential to influence when necessary.

- How could you break down the influence element of your work together? Find different parts of your work that you can apply your influence to. You cannot influence everything, but what can you influence? Sometimes the problem comes down to misunderstanding what each side can and should be able to influence, and what it is reasonable to expect each side to influence. For example, if your stakeholder is an external organisation, it will be appropriate for you to influence their response to your service requirements, but not their decisions on resource allocation or client strategy. You may wish to influence them to put your favourite project manager on your case, but that is probably not realistic or appropriate provided they deliver the agreed level of service.

- Consider this problem in the context of the wider group of stakeholders around your goal. Nobody works in isolation, and all are affected by a myriad of powerful others. There could be opportunities for you to leverage the relationships you have with other stakeholders to increase your influence with the stakeholder you are thinking of right now.

This is an important area of relationships that needs to be worked on consistently. There is little room for complacency. It builds on Trust and Credibility and, in turn, prepares the way for Problem Solving and Conflict Resolution.

Strengthening Problem Solving and Conflict Resolution

Usually, if you have managed to achieve high levels of Trust and Credibility and also Communication and Influence, this theme will look after itself. So this section is deliberately brief because I want you to focus on the first two themes to maximise your progress. Once you have strengthened them, you will have a really good stakeholder relationship. All that is needed here are a few supplementary points which build on this strong base and are relevant to this theme. In fact, they assume you already have a great relationship with your stakeholder.

- Problem Solving and Conflict Resolution is about cultivating and encouraging a robust attitude, which promotes proactive and objective attention to problems facing either party — hopefully, well ahead of them becoming a crisis. It is impossible to avoid issues arising in a relationship, but it is a deadly sin to leave problems lying around to fester.

- Care needs to be taken to avoid ascribing blame to people, making accusations and shaming people who you think have erred. Equally, watch your competitive spirit. While it can be great fun to add a little banter, if people feel embattled by you going into win mode, they are likely to forget collaboration and see if they can beat you instead.

- Establish clear processes in complex relationships that trigger a more sophisticated approach when a problem or conflict is identified. Normal dialogue and relationship management may be inappropriate when the temperature rises. If both sides can see the trigger, they can both recognise the need to adapt their behaviours and processes to best handle the problem or conflict well.

- Even in straightforward relationships, it will help to acknowledge the possibility that the usual flow of the interaction may alter. Agreeing what you would do if a serious disagreement arose can help to make it easier to adopt appropriate methods for dealing with a crisis in the relationship. Without this, there tends to be a lag between the need to change behaviour and the actual change taking place. And in that gap, bad feelings and harm can quickly accumulate. If you see a crisis on the horizon — get ready for it by working together.

- Watch out for the escalation. It is one thing to have a disagreement with someone who you usually get on well with, but if either side gets their superiors involved the whole game starts to change, rapidly. It is legitimate to involve others when a problem arises. Working with your stakeholder to make sure this is done in a way that maximises the chance you'll both come out with the best answer and the best relationship. How are you both going to escalate this appropriately? A great question for you both to discuss answers to.

- Maintain the warmth of your relationship while working on problems. Disagreement at work shouldn't prevent you having a beer afterwards. Just agree not to talk about the problems back at the office. By the same token, ensure you remember that you are dealing with humans, and where there are humans, there are feelings, sensitivities and vulnerabilities. Overly blunt words or direct feedback can easily arouse emotions that could detract from the process of fixing things.

- There is generally nothing wrong with admitting you've made a mistake that contributed to a problem, but it is wrong to avoid responsibility for trying to put things right. Okay, repeatedly making the same mistakes is wrong. Keep learning.

- Wherever possible, try to depersonalise the problem you are dealing with. That means adopting an objective stance where you can put the person to one side and look at the facts. And that person means you need to step aside from your emotions too.

- The point above may help to overcome what I often see as the biggest challenge with Problem Solving and Conflict Resolution — the natural desire for many people to avoid contentious situations. This can be a deep-seated personal defence system that is difficult to shift. No amount of process and objectivity can completely eradicate it. So recognise its presence if you have it, or if your stakeholder may have this inbuilt caution. Using high levels of assertion to force them to the table is unlikely to be the best way to engage them.

- Jointly recognise that the problem or conflict has come to an end. It's been concluded. Congratulate each other for prudent behaviour and arriving at a solution (even if you don't think you've won or come out ahead). Draw a line under it and move on. If it was a big conflict with lots of people involved, it might be a good idea to get the team(s) together to review learning and improve processes. This will make it easier and quicker next time a problem arises.

Key Points

- You are unlikely to be able to remedy relationship problems without a firm foundation of trust.

- You can have a brilliant relationship even when you don't agree all the time.

- Don't wait until a problem occurs before discussing how you will work on them. Any step towards agreeing resolution processes will improve the prospects of a swift solution. Problems are inevitable, so agree your process early.

- It is easy to revel in your power over others. If you do, beware of the kick under the belt.

- Reflecting on the themes is ideally done together because it makes improvement far more likely.

Suggested Actions

- Share this chapter with people who you want an excellent working relationship with.

- What action will you take now to improve the relationship you were assessing earlier in the chapter?

- Take action to reinforce the good aspects of your relationships too.

COFFEE BREAK

Being a Stakeholder

This book talks a great deal about stakeholders and mainly about your stakeholders. But have you stopped to consider that you are a stakeholder for others too? Where do you sit on the map of other people and projects around you? And, how are you behaving?

For instance:

o Are you disagreeing with what others want you to do, but not telling them why? It's a choice, but they may have labelled you as a Player. And if so, what sort of Player are you to them?

o Who is trying to strengthen their relationship with you at the moment? Why might this be the case? To what extent are you helping or hindering their attempt to get closer to you? Have you got that right?

o Who are the main people you are hindering at the moment? Why is that? What action might they be taking to isolate you, remove your objections, or simply push you out of the way? What's their strategy and who are their friends?

- For those you are agreeing with, are you supporting or really advocating them? They may not have the skill to influence you to do more for them, but you can volunteer greater advocacy.

- Your work may be causing negative consequences for other less powerful projects or goals. They may not be important to you, but they are important to someone else. Are there ways that you could lessen the damage you are causing to others?

This is all discretionary on your part, and you may be wondering, why bother? You have plenty else to do. All I am suggesting is that you at least consider the possibility that you could be building stronger relationships all around you without necessarily having to do a great deal more. Just open your mind to the possibility that you could become more helpful to others.

CHAPTER 12

Step 5: Engage

One of the temptations with the Stakeholder Influence Process is to fly into action as soon as you get the first big flash of inspiration. That may well yield the result you want, but please consider pausing just a moment longer. Make a clear decision about how you will interact with each stakeholder. You have your strategy and plan. The final part of the jigsaw is deciding the manner in which to approach each stakeholder.

This chapter will help you to:

- Consider how to pitch your idea, project or goal.
- Link what you want more closely with each stakeholder's agenda.
- Flex your behaviour to suit their personality.
- Decide how you're going to win them over.

If you're in a hurry:

- You probably won't suffer too much in the short term; however, a quick skim will make sure you are not missing anything critical.

The way you apply the ideas below will vary widely, depending on the goal you are working on and the stakeholder you need to influence. Some may have no relevance to what you are doing, but do not dismiss them too quickly. Instead, challenge yourself to work out how the idea could be adapted to add value in your situation.

Building a Compelling Vision

To gain influence, people have to change. An individual has to decide to make the change (albeit an often unconscious process), and this could be with reluctance or willingly. That means they have to be either forced or motivated to do what you want them to do or think what you want them to think. If your goal is big enough to warrant it, consider how you are creating a compelling and inspirational vision of the future for people to get excited about.

Once you've got a vision, you can use it as you engage with stakeholders. To maximise motivation to change, your vision needs to be:

- **Exciting**: With so many strategies walking the corridors around your organisation, where's the buzz that will make stakeholders sit up and take notice of yours? If you are struggling to get excited about your goal, why would anyone else?
- **Touchable**: Stakeholders are able to almost believe it has arrived. This is a difficult one, but unless people can believe it is capable of being realised,

they will be reluctant to put effort into helping making it a reality.

- **Logical:** Not normally a word you might associate with vision, but whatever vision you are putting out there needs to be credible and people need to believe it's a good thing to aim for, i.e. it fits in with everything else, or it's the appropriate way of disrupting the status quo.

- **Beneficial:** Along with stakeholders being able to almost touch your vision, they also need to feel they will fit into the new world and will be a beneficiary. While some altruistic individuals may accept their own demise, they are rather few and far between. And when I say benefit, I really mean that they will benefit more from the realisation of your vision than from any competing visions vying for their attention and buy-in.

Take some time to write down your vision. Aim to use as few words as possible. If you can get it onto a single page, so much the better. It can be used to show or send to people. Without a doubt, you need to have it in writing for your own benefit if not for anyone else. The act of writing is surprisingly powerful. It helps to straighten out your thinking and prepare you to deliver the message consistently.

If you are finding it difficult to develop a compelling vision, consult more widely about how others imagine the world would be like with your goal achieved. You can also start to imagine it from other people's perspectives. How might a customer describe the world with your goal realised? What about other teams within the organisation — how might they describe it to their friends?

Make sure to test your vision with a few close colleagues or friends and get their input on how to make it more exciting or compelling. They may not be as enthusiastic about it as you are, so don't let them discourage you from your ambition. Ensure you listen actively to what they have to say and critically appraise any amendments they are suggesting.

The bottom line is that if you cannot make a big goal compelling to stakeholders, you are likely to be faced with an ongoing struggle to influence them. You'll also have to keep influencing them to stay on your side. So, invest in some professional help. PR, Marketing or Communications teams could work wonders for you. Most people in large companies only consider using these people for their external customer-facing work, but everything they are skilled at can be deployed internally too. In my experience, they are usually more than happy to help people who ask.

Create a Benefits Register

If you have been able to get clarity in your own mind on what you are aiming to achieve, you will also have recognised all or most of the benefits you will realise from success. This is natural and to be expected, although it holds within it a risk that you may alienate your stakeholders.

Generally, stakeholders will not be too interested in how you are going to gain; instead, they want to know what they're going to get out of it (yes, just like you do). The degree of emotion and greed varies, but deep down everyone is considering the personal implications of any decision they make (remember the power principle of Calculations in Chapter 3?).

At some point in their deliberations, they will also be wondering what you're getting out of the deal. So when you are engaging with a stakeholder, make sure to be clear about what you're going to get rather than leave them to guess or fantasise.

To maximise your prospects of being able to motivate and engage stakeholders, carefully consider all of the benefits they will gain. This builds a resource of ideas for you that can be used with different stakeholders as you are influencing them. Although there will be common themes, there will also be wide differences between the motivations and hot buttons for different stakeholders — that is where this resource can come in very useful.

Coming up with benefits that others will gain that may involve a loss for you can also be useful. These can be extremely powerful persuaders if used carefully. Show others how much you are putting their interests ahead of your own. It also demonstrates that you have thought through the impact it will have on them, and they are likely to respect this too. But be careful not to push this too far. There is a fine balance between being credible in your care of other people and being perceived as either stupid or manipulative.

Another benefit you could get from this idea is that there may be things about your goal and work plan that could be adjusted in order to increase the benefits for others. Often, these changes can be made at little or no cost to you. So, when you're thinking this through, stretch further and look for opportunities to adapt what you are doing in order to increase the benefits for others.

A simple example of this came from a sales manager I was coaching a while back. She was struggling to get one of her consultants to buy into her new sales strategy. She had been trying to use the lure of all the extra commission he would earn if he hit his targets. What she had missed was that the primary motivator for this consultant wasn't money; it was time with his disabled son. When she adapted her approach to offer him an informal day off if he put his back into implementing the strategy, everything started to fly. He got his time off, and she got her revenue.

Another example of this was a senior manager I coached a number of years ago. He was finding it difficult to engage his Managing Director. He wanted to get his buy-in to a plan to recruit ten new sales staff. After we explored why he wanted ten, he realised that if it worked, the business could justify another 75 sales people to capitalise on the opportunity. So he began talking about his plan to increase the sales force by 50%, with an initial pilot of 10 new people. He soon had the Managing Director wanting to engage with him.

Being open about the gains and losses on both sides of the engagement is most likely to protect you from unhelpful suspicion and also maximise your credibility and influence. The best way to prepare for this is to create a benefits register in a similar way that you might create an issues log or a risk register in project management.

Tailoring Your Pitch

It is one of the most natural things in the world to be preoccupied with your own ideas and plans. If this spreads into the way you engage with stakeholders, it isn't going to get you very far. It is essential to translate your own ideas and goals into words and phrases that your stakeholder might naturally use, or that will speak to their agenda.

I'm a firm believer in the notion that you should have a consistent vision as the base for all communication with stakeholders. If you don't have this, there is a very real danger that when your stakeholders get together they may start talking about your goal and realise that they are getting different messages. Confusing a group of stakeholders like this is never a good idea and is likely to make you look vague and confused.

These two thoughts may appear like an either/or situation, but there is a very simple way through this dilemma. When you engage with a particular stakeholder, reference your vision and plan quickly and then say, "And what this means for you is ..." I'm sure you can think of any number of other linking statements that you could use. The important thing to notice here is that, because you are making reference to your vision, you are ensuring consistency, and then you are quickly making it relevant and interesting (hopefully) to the stakeholder you are talking to.

One Financial Controller I coached did this really well. He had just been appointed to his position. He crafted his vision and then began to implement it. He put the strapline for his vision in his email signature and added it to every document produced. When he attended each sub-team meeting to talk about the new vision, he simply wrote it on a flip chart and then began engaging them in understanding what it meant for them. This differed between the teams. He also used it at the beginning of each one-to-one, not only within his team, but elsewhere in the organisation. Simple idea, well executed.

Adopting Alternative Tactics

There has been a great deal of research done over the last few decades on influence that has yielded some fascinating insights into how to engage with your stakeholders. Dr Cecilia Falbe and her colleagues compiled and researched a range of distinct tactics that are commonly used in the workplace. They then set about considering the likelihood of each tactic being successful.

The great thing about their work is that it provides a quick checklist of different approaches you could use, so that you can decide which one is most suitable for your purpose. Provided you are aware of the likely consequences, you can potentially engage much more effectively with the right selection of tactics.

Inspirational Appeals

Here you seek commitment to your goals by appealing to your stakeholder's values, ideals and aspirations. This is directly related to the earlier sections of this chapter, as it is the behaviour built on your vision and benefits work. Unless you did that preparation, you are likely to be unconvincing if you try an inspiration appeal.

Consultation

The essence of this tactic is engaging your stakeholder in developing your detailed proposals or plans — before you've made up your own mind. Care is needed to avoid the suspicion that you are just going through the motions. A sincere inclusion in your decision-making process is a great tactic to get people onside before you've even started. Or, if you have already got moving, this tactic could involve you engaging them in problem solving.

Rational Persuasion

The use of logic and rationality is an extremely popular tactic when dealing with stakeholders but, there are limitations. The research has shown that, in actual fact, it is not the tactic most likely to succeed (I'll come back to this in a moment).

Ingratiation

In a nutshell, this tactic is about getting a stakeholder to like you so that they are more inclined to agree with you. Of course, everyone wants to be liked, or at least respected, but this specific tactic focuses the main influence attempt on being liked rather than rationality or inspiration.

Personal Appeals

Help! Often referred to as an emotional appeal, this is where you might try to call in a favour from a stakeholder, or simply beg them to do it. It plays heavily on the personal relationship, friendship and sense of loyalty.

Exchange

The subtle undertone of this is "you scratch my back, and I'll scratch yours" and if the parties have a good relationship, it is likely to be an explicit negotiation of terms. In an organisational setting, it could be a bargain struck with a stakeholder. If they support you on your goal, you'll withdraw your objections to their goal. Sometimes these exchanges are implied, with a nod and a wink.

Pressure

Assertion and aggression are effective influencing tactics, but are often criticised as being unfair or wrong. This applies less to assertion, but aggression is to be avoided for most people. The type of pressure applied can also vary, and as it changes, so does the common view of its acceptability. Pressure can include threats, bullying, nagging and public humiliation (either verbal or through email).

Legitimating

This tactic differs from Rational Persuasion because it seeks agreement in order to fit in with other organisational policies or procedures. There may be good logical reasons for retaining a poor performer, but this tactic uses the legislation as the reason to keep the individual. In many ways, this tactic is borrowing power from other sources.

Coalition

In the research, this tactic is referring the use of others to do your influencing for you. It is important to have other people on your side; however, excessive reliance on their influence is of limited benefit and risky in the long term. Unless you can win unaided, you will always be reliant to some extent on other people's power.

In their results, Falbe and her colleagues reported that the most successful tactics were Inspirational Appeals and Consultation. Somewhat surprisingly, Rational Persuasion languished in the middle in terms of success. Least likely to be successful were Pressure, Legitimating and Coalition.

The main surprise here is the position of Rational Persuasion. What they found was that when successful, rational argument is gaining compliance rather than commitment. The risk you face is that at best it compels people through logic rather than gaining their emotional buy-in. In order to gain commitment, you need to combine your logical argument with an Inspirational Appeal. Within many organisations, there is widespread over-reliance on rational persuasion.

It doesn't work anywhere like as well as most people think it does.

The other surprise is that Coalitions were in the least effective group. When you go deeper into the research, what becomes clear is that it is the specific use of a confederate to do your influencing for you that is less likely to be successful, partly because of the pressure it creates for the target. It does not mean that you should not build coalitions — these are extremely useful in getting more and more people on the right page and is really what the Stakeholder Influence Process is designed to achieve.

In order to bring these tactics to life, for each of your main stakeholders, consider:

- Which of the nine tactics have you tried before?
- Which were most successful for you with that stakeholder?
- Which have you not tried that could be useful?
- What tactics have you seen others use with this stakeholder? What happened?
- Which tactic(s) could you use to influence your stakeholder now?

Remember, there are no hard and fast rules here. What you need to do is adopt a tactic, or a range of tactics, which have been selected for a particular situation after careful thought. Blundering around with your favourite tactic will not be anywhere near as effective as selecting the right tools for the job in hand.

Adapting Your Style

The desire to influence other people is a natural part of being human. The way in which you influence (your style) has been established over the years by your experience and learning. Subconsciously, you will have found out what works for you. Yet each individual is different, and people will have found alternative ways to influence. Research I've conducted with colleagues over a number of years suggests a number of key principles relevant to influence and style:

1. Individuals differ in their influencing styles.
2. People prefer to be influenced in the way they like to influence others.
3. Differences in style create a distraction from the content of any communication.
4. Adapting style to remove the distraction creates stronger influence.

There is a simple message here. If you can become more aware of styles and then make clearer decisions about the style you adopt with each stakeholders, you are more likely to be more influential. The resources section at the end of the book will point you in the direction of more detail on this topic, but for now, here is a summary of the key areas of behaviour we have identified in our research:

- **Determination:** the preference to express clear views, opinions and goals and then drive them towards realisation vs. the preference to consult, accommodate and reach a harmonious solution, direction or view.

- **Tact and Diplomacy:** the preference to sense the feelings, concerns and agendas of other people and respond in a sensitive way vs. the preference to be direct and clear with others so they know where they stand, even if this risks upsetting them.

- **Sociability and Networking:** the preference to use social skills to build a wide and strong network of valuable contacts vs. the preference to focus on the task in hand and to avoid social distraction.

- **Emotional Control:** the preference to remain calm and focused on facts and process vs. the preference to express genuine emotions openly as they happen.

Although we have a psychometric instrument to assess these (The Gautrey Influence Profile is described in the resources section), you can get a rough feel by considering the extent to which you prefer behaviours that would match the first part of each dimension's description.

For instance, how strongly do you favour the use of social skills to build a wide and strong network of valuable contacts? Note that avoiding these behaviours leads to the second part of each dimension's description, i.e. focus on the task in hand and to avoid social distraction.

Think about each of the dimensions above and give yourself a score out of 15. If you score 15, you'll be strongly in favour of using behaviours that sit behind the first part of each description. If the second part of the statement is you all over, score yourself as 0. These are the score ranges the online questionnaire assesses.

Now comes the interesting part. How does this relate to engaging with your stakeholders? In a nutshell, if your stakeholder would score themselves differently than you score yourself, you've probably got some distraction creeping into your engagement. You may have the maturity and experience to move beyond this, but you may not, and they may not either. Let me give you a few examples:

- Highly sociable people love talking, often about themselves and what they did during the weekend (favouring Sociability and Networking). At the other end of the dimension, people won't be interested and may even resent the barrage of questions that have nothing to do with the task in hand. Their sociable colleagues will wonder what's wrong with them.

- If you are very calm and controlled (favouring Emotional Control), seeing huge emotional displays from others may be disturbing. You'll wish they could leave their feelings out of the debate. For their part, they will be wondering what you are thinking because they can't see you jumping up and down or banging the desk. They'll probably think you are a little aloof and dispassionate.

There are many other interesting dynamics at play between these dimensions that you can read more about in my other book, *Influential Leadership*. For now, this should be enough to give you some food for thought. However, it would be really useful to you if you pause for a few moments and think of a stakeholder you don't get along with that well, perhaps from the left hand side of your Stakeholder Influence Map.

- How do you think they would score on the dimensions?

- Notice how these scores differ from your own.

- Where are there big differences between the two of you?

- How does their behaviour make you feel?

- What do you think they feel about you?

- What implications are there for the way your relationship is/is not working?

- How could you behave differently so that the gap between you lessens?

Managing the Politics

Politics is an inevitable feature of organisational life. All of the various definitions lead to the behaviours people use when they seek to gain power and influence. These definitions are generally neutral when it comes to the intent or agenda behind the actions. Thus, people with high integrity can use the same behaviours as those of a more dubious disposition.

The actual difference in the way it plays out is in the level of deceit used and the harm caused to those affected. So, unless you work in an organisation where nobody is trying to influence or gain power, you'll need to come to terms with how politics works and how you can engage proactively (and hopefully authentically) so you can protect your goal.

Time does not permit us to do other than cover the basic principles, but the resources section at the end of the book will direct you to more specific coverage of this important topic.

To manage the politics, firstly you've got to see it and understand it. Knowing the tactics that people could use or are using is critical before you decide how to respond. One of my other books, *21 Dirty Tricks at Work*, helps people to understand reality by exploring political tactics such as:

- **Fall Guy/The Patsy**: Assigning projects or tasks that are destined to fail to an expendable manager so that they can be blamed for the failure, and/or to reassign favoured employees away from reputation threatening failure.

- **Rock and a Hard Place**: Manipulating people by offering limited or fixed choices, expecting the victim to choose the lesser of two evils.

- **Tell Me More**: The tactic of delaying decisions or honest disclosure by requesting more work, research or data which often includes the efforts of others.

- **My Hands are Tied**: Pretending to be helpless due to the influence of a higher authority or process, when under the same circumstances, but with a different person, there would be a different outcome; "Sorry, Ben, but the policy is ..."

There are many more in the book, and even more in the catalogue my co-author Mike Phipps assembled. However, the important thing to realise is that these are likely to be happening around your stakeholder community in various forms. So, try to think about other specific tactics which you see happening. This really helps build awareness and if you want to discuss it with a friend, so much the better.

You may also notice that these are somewhat tactical in nature. If you really want to get to grips with the bigger political strategies that people use (or fall into), have a good read of our other book *Political Dilemmas at Work*, written with Dr Gary Ranker. In here, you will learn about Consultants Rule, Home Alone, Power Vacuum and many more.

Once you've recognised exactly what is going on, the next step is deciding what to do about it. It is a complex subject which requires more than just a few pages to explore, so here are some ideas for you based on my experience of working with people of integrity, and helping them to succeed in highly tense political environments:

Step 5: Engage

- Keep your emotions under control, especially fear and paranoia. Try to look for the facts.

- Consider the big picture. Petty politics should be treated as such. Quite often, the worst thing you can do is lose yourself in their game playing.

- Offer ways forward/ways out. If someone is playing a trick on you, they are doing it for a reason. Try to find out what it is and construct a win-win proposal.

- Put out fires and build bridges. Relationships of trust are key and sometimes you need to be big enough to raise awareness and push for honesty.

- Walk in their shoes. They have different shoes, so try them out and see how the path feels for them. You may still not agree with them, but at least you'll be better placed to empathise.

- Take some time to think. Immediate reactions to political games rarely work, especially if you are not experienced at dealing with them.

- Ask them what they want. Why is this so rarely done? Don't shy away from asking the direct and straight question — just make sure to use a pleasant friendly tone rather than make it feel like an interrogation.

- Don't fight every battle. Whenever you meet a political trick, stand back a moment and decide if dealing with it will help or hinder your progress towards your goals.

Whatever you do, don't ignore or avoid the reality of political activity. If you do, it will catch you out — and probably at your most vulnerable moment.

Preparation for Engagement

So, now it is up to you to decide how best to engage with each stakeholder. There are no rights or wrongs, just that some approaches are more likely to get the result you want. It is also true that any engagement is probably better than no engagement. My advice to you now is to make some clear decisions about how you can engage, then see what happens and adjust as you go.

To help you to make your decisions, here are a few pointers. If you have been studying this book carefully, many will act as a reminder of ideas covered in previous chapters.

- Within the context of the strategy you have devised to achieve your influencing goal, select a stakeholder you want to engage with.

- What exactly do you want them to do, think or feel as a result of your influence attempt? Of course, you want them to buy-in to what you are doing but, what exactly do you want them to do as a result of that shift? Perhaps you want them to email their team demonstrating their support, or maybe, take up your case with their line-manager at their one-to-one next week. Get as specific as you can.

- Consider what influences this person. Look for examples of when you have seen them being influenced. What was happening? Why were they influenced? What was the influencer doing?

- Conversely, what doesn't influence this person? Can you think of examples when someone attempted to influence them, and it didn't work? What was tried and why didn't it work?

- What's in it for them? Standard sales technique here is to consider how your proposal will make life easier for them, or in some other way bring benefits to their doorstep. Linking your proposals to the benefits for them (not you) can have a dramatic effect on your prospects. Thinking about this before the meeting is critical.

- How will they lose? Every proposal has an element of disadvantage. You may be so entirely sold on what you are suggesting that you can see no reason why they could possibly say no. Fact is there will be reasons, even if these are not sufficient to thwart your attempts. These little seeds can be magnified quickly if something else upsets them as you attempt to influence them.

- How does your power compare with theirs? What sources of power could you use effectively with this person?

- How are you going to make your case? Structure your approach based on your earlier reflections and see if you can lay out a game plan. What arguments will you use? What evidence will you present? How will you reference wider support?

- Should you adapt your style of influence? If they are very different from you, be wary of the distraction that may creep in. What could you do to minimise that as you attempt to influence them?

- What about the indirect routes to influence? Are there easier ways to get what you want?

- Are you the best person for the job? I hope you are, but don't forget to think of people in your team and elsewhere in the organisation who you might delegate this to.

- Where you attempt your influence may also make a difference. Their office, yours, or someplace else?

- Do you need to bring along other people to get the job done?

- How might they respond to your attempt at influence? What possibilities are there and what can you do to reduce the risks or prepare for those eventualities? Don't think too hard about this. Otherwise, you'll wrap yourself up in endless possibilities with low probability. Sometimes you've just got to go and get on with it.

- Speaking of which, when will you make the attempt? Timing can often make or break the attempt.

If successful influence is really important to you, consider all of the points above. I know there is a lot but by now, if you've been reading this book carefully, most of the answers will flow quite quickly. With a little practice, you'll be able to do this while walking to the next meeting.

Key Points

- Influence is easy when they see how much they are going to benefit from what you are doing.

- Vision and inspiration are not just for CEOs and top leaders. Every great influencer makes use of these techniques. So should you.

- Most people use habitual tactics rather than the ones that are most appropriate to the individual and the situation.

- You cannot avoid politics, but you don't need to get immersed in negative politicking.
- Adapting your style is simply a technique to remove or lessen unhelpful distractions.

Suggested Actions

- Observe the styles that those closest to you adopt, study the differences and practice changing in safe situations.
- Become a student of how others are attempting to influence you. You may learn some really useful ideas.
- Make decisions and start to execute your plan and achieve your goal.

COFFEE BREAK

Remote Team Members

With the increasing use of global teams to implement projects, the chances are growing that you will be responsible for team members who you may rarely meet, if ever. Traditionally, it was taken for granted that you would have a face-to-face relationship. Many project managers are missing this and remain stuck in the past. That is not wise because the world is not going to turn back.

Here are some thoughts to reflect on while you're having a coffee:

- Remote relationships can work exceptionally well if both parties believe it can and are willing to put the effort in.

- Personal/social relationships are just as feasible remotely as face-to-face, but you have to invest some time in making it happen.

- Rather than ignore this aspect of your project, raise it on a regular basis and get the whole team talking about how to make it work more effectively.

- Take time to review (and discuss) the significant benefits remote working brings to the team.
- Make sure to match the medium with the task in hand. Sometimes calls are better than emails and vice versa.
- What can you do right now to improve the quality of your remote relationships?
- What can you do right now to help your team to work more effectively in a remote way?

Personally, I'd much sooner work remotely than spend my time struggling through the traffic or hanging around in airports. It's all about frame of mind and belief. So, don't let anyone get away with the remote working excuse for poor performance.

CHAPTER 13

Step 6: Maintain

Great as the ideas in this book are, they are almost worthless if you don't create and maintain a habit of applying the process. It should be used with all of your major influencing goals and reviewing your progress towards each on a regular basis is essential.

This chapter will help you to:

- Periodically review progress towards your goals using the Stakeholder Influence Process.
- Cultivate increasing levels of personal motivation.
- Maintain momentum towards your goals.

If you're in a hurry:

- To be blunt, if you are in too much of a hurry to review your progress, it won't be long before your lack of progress becomes obvious. Just use this chapter as a checklist for how to review your progress thoroughly. It won't take long.

If you want to keep moving towards your goal, you have to keep motivated to do whatever it takes. Few great achievements are accomplished without persistent effort to keep moving forward. To avoid taking wrong turns as you move, you need to keep refreshing your plans and tactics, learning from experience and adjusting your actions. At the same time, you can also attend to keeping your motivation in peak condition.

In this busy life, there are far more things clamouring for your time than you could possibly attend to, so you'll need to find a way to make sure that your influencing goals remain high on your list of priorities. If you have chosen your goals well, this should be fairly easy. The ideas below can help make sure they remain prominent, and help you to achieve more than you originally intended.

Increasing Motivation

The main principle here is that you need to find a way to keep going — or perhaps, many ways to keep going. If you've got this far in the book, you're already demonstrating motivation, but where will this energy and enthusiasm be in a couple of months' time? When the going gets a little rough, you need to be ready — you need to be at your peak in motivation terms, so that you have the tenacity to stick with it.

Since motivation is such an important factor, you may wonder why it comes towards the end of this book. I have presumed that your motivation was high at the beginning, so it wasn't necessary to put it right up front. Now that you are nearly at the end of the book, this section can be used as a little preparation for when the times get tougher and your motivation wanes.

One way to do this is to keep the personal benefits that success will bring to you, and also the negative consequences or losses you will incur if you fail, at the forefront of your mind. Some people tend to be more motivated by the gains they could make, others by the losses. In psychology, this is sometimes referred to as "moving towards" or "moving away" motivation. Whichever you prefer, that's okay — recognising it will help you to focus your thinking in the next action.

Motivation Exercise

Get out your Stakeholder Influence Map. On the back, draw a vertical line down the middle. On the left of this line, write down everything you will gain from achieving the goal you have been working on. On the right, note everything you could lose if you fail. Refer to this often, but especially whenever you are doing a progress review of the Stakeholder Influence Process.

For instance:

- How will it contribute to your pay review or bonus?
- Will it improve your career? How?
- Does success with this goal raise your profile?
- Will it make life easier for you?
- How will others you care about benefit?
- What problems will it solve for you?
- How will it improve the way your colleagues think about you?
- What new connections/friends will you make?
- Will it improve your personal life?

- What will you lose if you fail?
- How will your reputation be damaged?
- Will others think you have let them down?
- Could you get fired or rejected for promotion?
- How will more senior colleagues think about your failure?
- Will you have enough excuses to get away with it? Will anybody buy them?
- What will your partner or friends think?

This is very similar to the suggestion in the previous chapter where I recommended compiling a benefits register. This time it's personal.

Thinking about the gains and losses for your goal is great, but being realistic, you also need to factor in all of the other projects you have on the go. Your current goal may have lots of potential benefits, what about goal B or even goal C? Perhaps goal A is worth ditching in favour of goal B. Only you can decide, but decide you must. As soon as you can, complete the exercise for each of your major goals.

A good way of pulling these all together (once you've done the exercise above for each) is to summarise them all on one sheet of paper. From your individual lists, summarise the key benefits of achieving the goal and the pitfalls if you fail.

When you look at them side by side, it may be time to de-prioritise some of the goals you are working towards. If you do, don't forget to manage the expectations of those stakeholders you are going to be upsetting by not working on something they are interested in.

Exceptional Motivation

Now, how about raising the stakes a little more? They say that fortune favours the brave, so consider these ideas to increase your commitment and motivation, or even fear if that's what gets you going.

- Have a bet/wager with a competing project manager that you will win the resources/get the budget/be first to complete.

- Make public statements about the deadlines you have set.

- Strike a bargain with one of your stakeholders; with consequences if you fail to deliver (you could have a bit of fun with this too).

- Use every opportunity to stress just how critical your goal is to the success of the business (careful now).

- Get your project on the radar for the compliance/governance committee.

- Talk up the consequences of failure for the organisation and everybody who works there.

- Tell your boss you will resign if you fail. (Getting nervous?)

- Find ways to boast to your friends about how important your project is.

Be careful you don't back yourself into a corner and become too personally attached to your goal. At some stage in your regular reviews, you may conclude that you need to force the closure of your own project. It would also be wise not to alienate your colleagues and friends with these approaches. Err on the side of light-hearted banter and cultivate a bit of fun.

Okay, I will confess; some of these suggestions may be a bit provocative, but that's my job. You need to consider carefully all options that could keep you on the move because of the serious (positive) consequences for you and your career when you achieve your goal.

A final idea about motivation for you is to use your stakeholders to keep you motivated. Advocates and Enemies can be quite useful here. By reminding your Advocates of the benefits they will accrue, you can enlist their guidance on how to keep you moving forward. Perhaps they could exert a little discipline or regularly ask for updates. On the other hand, reminding yourself of all that your Enemies could gain from your downfall could give you just the boost you need.

Regular Review and Refreshment

The Stakeholder Influence Process is not a one-off event. It should be reviewed and regularly refreshed until you have achieved your goal, or until the goal is no longer critical for your valuable time and attention.

There are two main ways to do a review:

- ○ A Key Question Review where you review your progress, learning, and insights before deciding what to do next.

- ○ A Process Step Review where you take a fresh look at one of the steps in the process to find creative new opportunities for advancing your goal.

Each type of review requires a slightly different frame of mind; so if you are tempted to do them both in a single session, try to put a clear structure in place. Otherwise, you are likely to lose momentum and wander around the topics unnecessarily. Do the Key Question

Review first, then stop, take a break and then do the Process Step Review. At the end of the session, pull together your ideas from both types of review when deciding how to refresh your strategy or refocus your action.

Whichever review you do, make sure to have a copy of your Stakeholder Influence Map in front of you, or better still, on the wall so the whole team can see it.

Key Question Review

Here you focus on what has been happening, what you have been learning and what you need to do differently to accelerate progress during the next period of implementation. These questions will quickly help you to find new ways to boost your progress:

1. What progress have you made?
2. What has changed?
3. What have you learned?
4. What action can you take now?
5. When will you review again?

You should also stop and consider if you are actually focusing on the areas that will give you the greatest progress. Remember Pareto and his 80/20 rule — 80% of your results come from 20% of your efforts (and sadly, the opposite is also true). So, keep challenging yourself to look for the 20% that will give you the greatest progress towards your goal.

Process Step Review

For this type of review, remind yourself of the steps in the process of the Stakeholder Influence Process and choose one step to focus your review on. It could be that you notice one step that is causing you problems in your progress, or could hold strong opportunities for rapid acceleration. If so, focus your review there by reminding yourself of the key points in the relevant chapter. Also, consider the points below specific to the step you are reviewing.

Here are the steps in the Stakeholder Influence Process with additional ideas pertinent to the review process:

Step 1: Focus. Assess your priorities and focus your influencing goal

After a period of implementation, you may start to realise that the goal you have chosen, or the way you have articulated it, is not as helpful as it could be. Just because last week you thought it was the right thing to shoot for, it doesn't mean that it has to stay that way. If during the week you have discovered new intelligence in the political side of your organisation, it may be an extremely wise decision to amend your goal. For instance, if you have discovered that the most powerful person in the organisation is going to lose out heavily if you achieve your goal, it would be almost mad to continue pushing for it!

More likely, you could have discovered that a key stakeholder is working on something similar. If you re-position the wording and the direction of your goal to align more strongly with theirs, you are likely to be able to ride on the back of their influence, as well as help them to move forward with their goal, thereby creating coalitions and allies.

Step 2: Identify. Work out which stakeholders can have the biggest impact

It is surprising how quickly your knowledge of the organisation will improve once you get going with the Stakeholder Influence Process. Once people have grasped the concepts of organisational power, they start to look out on a different world, noticing things that before would have been insignificant to them. If you are including other close associates in your review, the pooling of your intelligence will magnify the learning for everyone.

Consequently, there are often significant changes in the stakeholders named on the map during the first couple of cycles of the process. That's okay and natural. Each time you will be getting more effective and closer to hitting the right buttons to achieve your goal.

Step 3: Analyse. Map the position of each stakeholder

Given that the actual purpose of the Stakeholder Influence Process is to shift sufficient power and impact into the Advocates box so that you achieve your goal, things should be changing here all the time. Expect to redraw the map many times during the pursuit of your goal. They tend to get very untidy quickly, and that is not helpful when you are engaging others in your thinking and planning. But be warned, if you haven't got it written down, you are not making effective use of the process. So relax your natural attention to neatness and scribble all over them. That you have to redraw them is a great way of reviewing where everyone has moved to in any case.

Step 4: Plan. Decide your strategy for increasing buy-in

Without doubt, one of the most striking things that often occurs during the process is the realisation that you are pushing in the wrong direction. Sometimes you notice new opportunities for collaboration with other projects or join forces with someone else. At other times, you may be struck by the realisation that you have misinterpreted some of the fundamental aspects of your project and its fit within the context of the organisation.

These insights can come during any step in the first iteration of the Stakeholder Influence Process, but it is more usual for them to be exposed during a review.

Step 5: Engage. Adapt your approach to influence your stakeholders

Despite the guidance given elsewhere in this book and the huge amount of skill you wield, engagement is a learning or evolving process — there is always room for improvement. If your progress seems to be stuck, chances are high that the problem (and the solution) could lie in the way you are engaging with your stakeholders. If you think something is not working, but can't tell what, make sure to get some input from friends, Advocates and Critics. Because of your good relationship with them, they should be quite happy to give you feedback and help you see things in a different light.

In closing this chapter, I cannot stress highly enough the benefit of including others in your deliberations. Nobody can see everything, and ambitious people are particularly prone to closing down alternative views as they focus more determination on their goal. If you are working with a team, get them up to speed with the Stakeholder Influence Process. Share it with your boss and even your stakeholders. The more people you can include in your review, the richer you will become; richer in insight, action and probably financial rewards too.

Key Points

- If you are working on big important goals, your motivation and commitment will come under attack. It is essential to be proactive in preparing for this inevitability.

- Although motivation comes from inside, you can establish external mechanisms to keep it high.

- Using the questions and the processes in this chapter will ensure you do a rigorous review and maximise learning and potential for progress.

Suggested Actions

- Diarise your reviews so that they don't get forgotten.

- Get input for your reviews from others around you, especially stakeholders.

CHAPTER 14

The Stakeholder Influence Process

Here's how to get to grips with the whole process in just an hour. I will be taking shortcuts, and you will need to remain alert to when you need to delve into more detail elsewhere in the book. If you've been reading the book sequentially, this will be a great summary of the whole process to reinforce your learning.

This chapter will help you to:

- Gain an overall view of the process and how to use it.
- Get moving fast and try the whole process out.
- Make immediate progress on something that is important to you right now.
- Consolidate your learning so far.

If you're in a hurry:

- This is your first chapter. Read it fully.

The Stakeholder Influence Process

```
        ┌─────────┐
        │    1    │
        │  Focus  │
   ┌────┴─────┬───┴─────┐
   │    6     │    2    │
   │ Maintain │ Identify│
   ├──────────┼─────────┤
   │    5     │    3    │
   │  Engage  │ Analyse │
   └────┬─────┴────┬────┘
        │    4     │
        │   Plan   │
        └──────────┘
```

Step 1 – **Focus**: Assess your priorities and focus your Influencing Goal.

Step 2 – **Identify**: Work out which stakeholders can have the biggest impact.

Step 3 – **Analyse**: Map the position of each stakeholder.

Step 4 – **Plan**: Decide your strategy for increasing buy-in.

Step 5 – **Engage**: Adapt your approach to influence your stakeholders.

Step 6 – **Maintain**: Keep the momentum going with regular reviews.

Any process aiming at influence needs to quickly move towards action. Only through action will you really accelerate your result. You don't need days of learning to figure out how to apply it — you can start right now with this chapter, and then come back later and go deeper into the other chapters to refine and build your practice. If you have been diligently reading through each chapter, this one will summarise all the key points for you so you can get moving — if you haven't done that already.

This chapter contains the absolute minimum to start applying the Stakeholder Influence Process to your work. It will guide you through the process and help you to quickly build a strategy that you can then start to implement.

My expectation is that within one hour you will have discovered some new actions that you can take to improve your prospects and start to move towards your goal. As I have said before, this process will help you to find the actions needed, but does not provide the solution to all your problems. After working with thousands of people learning to use the Stakeholder Influence Process, I can assure you that what you will almost certainly find is that it helps — big time. All you need to do is apply yourself well over the next hour, and you should be ready to make some fast progress.

If this is your first time through the process, try to resist the temptation to wander around the other chapters — keep focused on the steps outlined here.

Firstly, take a quick look at the diagram of the process below to get a visual overview of the whole Stakeholder Influence Process. It starts at the top and progresses clockwise through the other steps.

Step 1: Focus. Assess your priorities and focus your influencing goal

The starting point is working out what you want to achieve. Sure, you will have lots of goals, but you need to get focused in order to gain maximum progress with the Stakeholder Influence Process.

Pause and think of the most important goal you have right now.

I am sure you have lots of goals, but which one is most prominent in your mind right now? Settle on one to use over the next hour for your first trip through the Stakeholder Influence Process.

To achieve your goal, what have you got to influence to maximise the probability that you will succeed?

This is often slightly different to your actual goal. If you want to achieve a 5% market share for your product within the next year (your key goal), the most important thing you have to influence could be that the board buys into your strategy and gives you the funding you need. Alternatively, the key influence may be to get the legal team to agree that it is prudent to implement your project.

Your answer to this question will help you to establish an influencing goal that is orientated towards creating a change of some sort in a defined group of people or an individual. This will now be your focus for the Stakeholder Influence Process.

You'll have lots of things you need to influence for each goal, and likely many different goals. So keep it simple and focus on just one right now; one that you think is really important to your progress. Later, you can come back and do the process again for other challenges and goals.

Chapter 3 goes into much more detail about focusing your influencing work — take a look when you have a little time on your hands.

Step 2: Identify. Work out which stakeholders can have the biggest impact

Okay, so if that's what you want to influence, what you want to make happen, who can help or hinder your progress toward achieving that influence? Think about all of the powerful people who can have an impact, both for and against. These people will become your stakeholders.

Brainstorm the names of all key people, both close to your goal and those who have a vested interest.

It is vital to stay focused on your influencing goal. Don't wander off to people like your career stakeholders (unless that is your focus). Instead, keep them all relevant to your goal.

Write down a list of eight to twelve people who can have a big impact on your success.

Don't worry if the names you are coming up with are difficult to engage with. If they could have a big impact, put them on the list and worry about how to influence them later.

Although Chapter 5 is the main place to go to in order to explore this step in more detail, Chapter 4 will be very useful in stimulating your ideas on who should really be on your list of stakeholders. It explores the fascinating subject of power and how it shapes organisations and the decisions they take.

Step 3: Analyse. Map the position of each stakeholder

Once you've settled on an initial list of eight to twelve stakeholders, you can now start to consider their position on the stakeholder map below. Take a look at the map and read the notes that follow. Then draw it out in your notebook and plot your stakeholders.

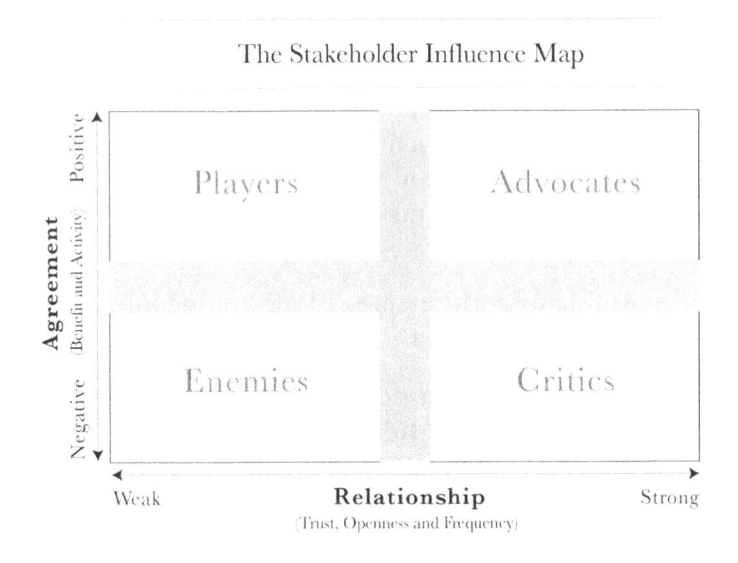

The Stakeholder Influence Map

```
                    Positive ▲
                             |
              Players        |        Advocates
Agreement                    |
(Benefit and Activity)       |
              Enemies        |        Critics
                             |
                    Negative ▼
        ◄────────────────────┼────────────────────►
        Weak            Relationship            Strong
                  (Trust, Openness and Frequency)
```

Where do you think each of your stakeholders is positioned on these two dimensions?

Relationship

- **Trust:** Do you trust them and do they trust you?

- **Openness:** Would they volunteer information if they thought it could help you even if you didn't ask for it? Would you?

- **Frequency:** Do you interact/engage with them often?

Consider these factors and make an initial decision about where to position them on the horizontal dimension of relationship. If you've got a great relationship with them, they'll be heading toward the right side of the map. However, if you have had some bad experiences or the evidence is somewhat patchy, maybe they'll fit into the left of the diagram. If you are really unsure, perhaps because you don't know them very well, leave them in the grey zone — that's okay for now.

Agreement

- **Interest:** Will they benefit or lose if you are successful? By how much?
- **Agreement:** Do they agree with what you are trying to achieve? They may lose a great deal but still agree that it is the right thing to do, for the benefit of the organisation as a whole. The opposite could also be true.
- **Activity:** To what extent are they actively supporting or blocking you? This is often an indicator of the way they are thinking in terms of agreement and/or interest.

What you need to do is arrive at an initial position for them on the vertical dimension of agreement. If they are clearly in favour of what you are doing, seem to be a beneficiary and have demonstrated their support, they're heading for the top half of the map.

Now, quickly take each individual on your list and write their names in the appropriate position on the map in your notebook. Don't agonise over it, think about it and write their name in a box. This will be your initial positioning to review later, maybe after you've taken some action.

You may find it helpful to pause a moment in order to understand the different boxes. The labels are chosen to deliberately provoke good thinking, not to throw accusations about. So to put a little more meat on the bones:

- An **Advocate** is a fan. Someone who really believes in what you are doing and is willing to put themselves out in order to help you succeed.

- **Critics** are people who you trust and will tell you what's on their mind. You will believe what they say. They will honestly point out the flaws in your arguments and will also be open to negotiation. You know where you stand with these people.

- **Players** are the sort of people you never quite know what they really think. They seem to say all the right things, but never seem to follow through. In a meeting, they are likely to agree, but their actions speak louder than words.

- **Enemies** — well, okay, in most cases this is pushing it a bit, but generally these are the people who you don't get on well with and are quite happy for you to know that they don't agree with what you are doing. Sadly, they will never quite level with you about what they are going to do to stop you.

- People in the grey areas are the ones who you are unsure about. Later, you will be taking different actions depending on which boxes they fall between. If they are people with power, they will need to be taken seriously.

There is much more detail on the things you need to be considering on this step in Chapter 7. However, this extra detail is only useful right now if you are really stuck — I would much sooner you have a go and move on to the next step. Leave the deeper immersion in the application of the Stakeholder Influence Process for your first review.

Step 4: Plan. Decide your strategy for increasing buy-in

So what do you need to change? When you have placed the impactful stakeholders on your map, you need to start thinking about who you need to move, and where.

Quite often, there will be one individual who holds the key to bringing all the others around to your way of thinking — or to your opponent's way of thinking. These people may not be the most obvious. In fact, they are often a few steps removed from the distracting cut and thrust — yet they exert massive influence over the way things are moving.

Pause and consider what needs to change to improve the situation dramatically

Try to come up with three key things which need to change. These questions will help your thinking:

- What one change would create a massive movement towards your goal?
- Who is the key person blocking you right now?
- What core attitudes need to shift to remove all obstacles?
- If you could wave a magic wand and move one stakeholder over to your side — who would that be?

Try to focus on the bigger picture and the things that, if you could change, would have a snowball effect — the things that would sweep away opposition and make success a foregone conclusion. Overall, your aim should be to move sufficiently powerful (impactful) people right and upwards to achieve the influence you desire — and make the accomplishment of your goal a certainty.

Chapters 6 to 10 contribute greatly towards developing an effective strategy. To be honest, on most occasions when I am taking clients through this stage, they don't need much help beyond the questions above to figure out who they need to be working on. What those chapters will do is help to make sure that you are finding the best strategy in your situation rather than the first to come into your mind.

Step 5: Engage. Adapt your approach to influence your stakeholders

Chapters 9, 11 and 12 go into more detail about how to engage with each type of stakeholder, but you don't need that now. If you are like most of my clients and workshop delegates, the preceding steps will have already opened your eyes to what I often call a "blinding flash" that has been sitting in front of you, begging for attention. So, at this stage, I'll just give you a few pointers:

- Go seek the advice of your Advocates. They are your best friends and can (and probably will) give you wise counsel on how to solve the challenges you have elsewhere on your map — particularly with Enemies.

- Engage your Critics with a positive attitude. You have a good relationship and can take a negotiating approach with them. The great thing about Critics is that they will be honest with you, so you can find out what you need to do to win them over.

- If you need to, aim to build your relationship with Players. There is something not quite right about the way you two are communicating and working together. So, consider bringing this to the table and (without threatening) see if you can cultivate a more open way of doing business together. Alternatively, work with more powerful people, so the Players don't cause you any problems.

- ○ Don't worry too much about your Enemies. Of course, you have to do something about really powerful ones, but generally I find that these characters revel in the attention. Instead, work with Advocates (and even Critics) to minimise the damage these people could do.

Later, when you have time, you can also take a look at Chapters 11 and 12 about engaging stakeholders and strengthening relationships. But don't worry about that now unless you absolutely have to. Generally, once people work out who they need to focus on they also find they are more than capable of making it happen. At risk of appearing dismissive of my eminent colleagues in other disciplines, a healthy dose of common sense doesn't need a liberal amount of theoretical confusion.

Step 6: Maintain. Keep motivated, moving and refreshing

At this stage, this is pretty easy. You're already motivated and moving, so fix a time in your diary to refresh your plans by reviewing your progress. You've got to keep coming back to this process to review how you are doing, what you have learnt and what else you need to do to keep moving towards your goal.

Chapter 13 provides ideas and processes for your review, but before you look there — do something first. Take action to start moving faster towards your goal and then take a look at Chapter 13.

My hope is that this hour has been well spent, and you have come up with some great ideas for moving forward your results with greater buy-in from your stakeholders. I also hope that you have been able to get a feel for the potential of the Stakeholder Influence

Process. It really is quite simple once you've understood the basics, but it is also quite subtle, and deeper understanding and continued learning will reward you well.

If you need it, the next chapter will take you through a simple example based on a real client situation. After that, it's just a question of you getting on with using the process and keeping coming back to strengthen your practice.

Key Points

- The purpose of this process is to guide your thinking quickly and move you towards action.
- Within an hour, you can develop your thinking dramatically. Just stick to the process.
- Don't over complicate the process. Learn to take action based on imperfect information. As you go, stay alert to learning and adjusting your action.

Suggested Actions

- Just do it.
- Come back later and use the other chapters to learn more.

CHAPTER 15

Project Hawaii

My sincere hope is that you have already had a go at the full Stakeholder Influence Process by now. If you have, this will be a quick read. More a case of confirming that you've understood the main ideas. If you haven't yet had a go with the process, I hope that this chapter will give you the final insights necessary to get going.

This chapter will help you to:

- ◦ Relate your learning so far to a practical example so that you can see how each element of the process works.

- ◦ Challenge your thinking by looking at the problems and challenges someone else has faced.

- ◦ Confirm by example the practical benefits you can gain with the Stakeholder Influence Process.

If you're in a hurry:

- ○ Don't worry about this chapter if you have already convinced yourself that you can make use of the Stakeholder Influence Process.

Project Hawaii: An Example

To illustrate the practical nature of the Stakeholder Influence Process, what follows is an outline of the main problems facing Jim, a former client. To help you understand how to implement the process, it is only necessary for me to illustrate the first four steps. It would be fun to speculate on how he should engage with his stakeholders, but I don't believe that would be a valuable use of your time.

While Project Hawaii is based on a real client, naturally, the names and the situation have been disguised.

Step 1: Focus. Assess your priorities and focus your influencing goal

As a busy programme manager, Jim had plenty of things to do. He had five different projects on the go, most of which were going along reasonably well. The one that was worrying him the most was Project Hawaii. The name disguised the boring nature of this attempt to innovate the company's Management Information System. He was struggling to get the resources he needed, and some key deadlines were looming.

"If only Anne, my sponsor, would get a little more active and help me get the resources I need from Finance."

This had been occupying quite a bit of time, and so far Jim had failed to make any progress on this. He noted down his influencing goal as:

"Finance will provide 50 hours of suitably qualified resource per month (at least grade C10)."

Influencing goals can always be specified with greater accuracy, but don't get overly worried about it. Provided it's good enough to get you focused, move on. You can always refine it during a review

Jim could have set the goal as *"Anne will get Finance to ..."* but this would have been less useful as a focus because it only involves influencing one person (look at Chapter 3 for more on this).

Step 2: Identify. Work out which stakeholders can have the biggest impact

As Management Services Director and sponsor of Jim's project, Anne was obviously a key stakeholder. Jim also noted down the existing project team members who were feeling the strain because of the gap in resources. The key ones were Peter, Sanjay and Felicity. Clearly Marco, the Finance Director, could have a big impact — particularly since he was the one currently saying no. After a little more thought, it occurred to him that the people who would benefit most if his project landed was the Managing Director and the two Operations Directors. They would use the real-time data to make their day-to-day decisions as they grew the business — so he added Joe, Dawn and Bernd.

Luckily, Jim was able to chat it through with a close colleague, who asked quite a few challenging questions. One revelation was that Marco was facing requests for his resource from several other projects too. There were two big ones:

- ○ Firstly, Project Malta run by David, would launch a new product onto the market. This project was sponsored by the Marketing Director, Tanja. Rooting for this was Sally, Head of Sales. She was quite active drumming up enthusiasm for this big opportunity.

- Also, a little known project looking at the budgeting process. This had been going on for some time, reporting to the Finance Director. Like many of these types of projects, it was resource hungry and Charlie, the project manager, seemed to be quite good at holding on to the resources he needed.

There were lots of other people, but these seemed to be the main ones who could help or hinder Jim in securing the resources he needed to complete his project on time. In summary, the stakeholders were:

- Anne: Sponsor of Jim's project.
- Peter, Sanjay and Felicity: Jim's key project team members.
- Marco: Finance Director.
- Joe: Managing Director.
- Dawn and Bernd: Operations Directors.
- Tanja: Marketing Director.
- Sally: Head of Sales.
- Charlie: Project Manager for Budget Process Control.

Step 3: Analyse. Map the position of each stakeholder

What Jim realised when doing the analysis:

1. Although Sanjay was on his own team, Jim noted that he was often making supportive noises, but was repeatedly failing to play his part in pushing for more resource. Jim started to suspect that he could be a little sensitive about his role on the

project and may be feeling vulnerable. If a more experienced resource joined the team, Sanjay could feel threatened. It was only a hunch, but there was definitely something not quite right.

2. Anne was an Advocate, but that title didn't quite fit. She should be very much in agreement, but why had she not been able to make it happen and get Marco to allocate the necessary resource? They had always got on well, and Jim had noticed lately that she seemed a little distracted or at least disinterested in Hawaii.

3. Marco was clearly against giving up his resource; otherwise, it would have happened by now. Jim didn't know him very well. There had never been any trouble or reason to doubt what he said — they'd nod to each other in the corridor, but that was about it.

4. Tanja was a completely different story — she was always fun to be around. However, behind the smile, it was clear she wanted to get her work done first and had even joked over a coffee that Jim would just have to wait his turn.

5. Charlie had a fearsome reputation. He was one of those devious characters who seemed to delight in seeing others struggle. It had even been rumoured that he'd gone out of his way last year to embarrass someone on the board for no other reason than to have a bit of fun. He did once try that on Jim, but didn't quite get away with it because the MD (Joe) took him to one side and had one of those little chats that were part of company folklore. Of course, Joe was on the same page when it came to Hawaii, but Charlie, no way.

6. Dawn and Bernd were the types that always said the right thing, but never really seemed to do

anything. Only last week, Dawn had agreed to put pressure on Marco to get the resources sorted out, but nothing had happened — and not for the first time. He had started to think that they were just making the right noises to keep on the right side of Jim.

7. Sally was a bit of an enigma. He couldn't quite figure her out. In fact, he wondered whether she should be on the stakeholder map at all.

Below is the stakeholder map Jim produced.

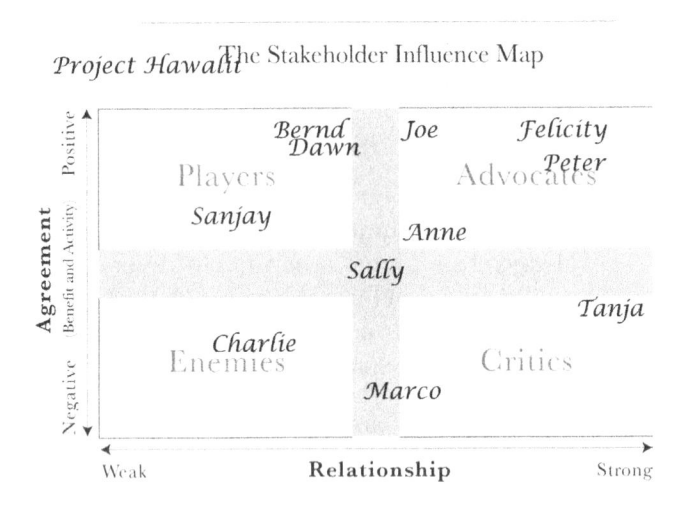

Project Hawaii The Stakeholder Influence Map

```
                   Bernd        Joe         Felicity
                   Dawn                     Peter
         Players               Advocates
                Sanjay
                              Anne
                       Sally
                                           Tanja
                 Charlie
         Enemies                 Critics
                        Marco

   Weak            Relationship            Strong
```

(Y-axis: Agreement — Positive / Benefit and Anxiety / Negative)

Step 4: Plan. Decide your strategy for increasing buy-in

To cut a long story short, Jim realised that Sally and Tanja were becoming much more powerful. They had been highly successful lately in working together to bring new products to market and had made the company lots of money. Marco was someone who had been powerful for a long time. But after many rounds of cost-cutting and dwindling market share, many were starting to think that the cuts had been too deep. Last year's budget round had seen the marketing/sales budget rise by 20%, which came as a big surprise to many people around the office.

Jim had not considered the link with sales before — he was really a numbers man and had been concentrating on getting the system delivered so that the MD could make decisions. Of course, one of the benefits of the system was that sales numbers were easier to get, and that could help Sally to adjust her sales strategy and tactics much more quickly. Her traditional approach was gut feeling, which seemed to be doing okay. However, if she could back it up with numbers, she'd probably be able to capture even more sales and get an even bigger development budget next year.

So his strategy headlines became:

1. Convince Tanja and Sally of the benefits of having immediate sales data (i.e. move them into the Advocates box).

2. Then ask them to help him get the resources he needed from Marco to complete the project and get the system ready for use when they launched their new product.

3. Get closer to Marco and build more of a personal relationship to make things a little warmer.

4. Put in place a standard communication plan to keep all stakeholders up to date on progress, but also regularly reinforce the commercial benefits that would flow from full implementation.

To put this more simply:

○ Motivate Tanja and Sally to get resources.

- Make friends with Marco.
- Implement Communication Plan.

One of the great benefits of the Stakeholder Influence Process is its flexibility. It can handle whatever level of detail you want to throw at it. Provided your goal is close to the criteria set out in Chapter 3, it should help you move forward. Of course, you can always go deeper, and that is very tempting when you are using the map with other colleagues, but I strongly recommend that you move quickly to action. There is little point in spending hours and hours pursuing the ultimate answer. The only way you can arrive at that is if you have perfect knowledge. Since most of the time you need to be probing into people's feelings, attitudes and power, personally I think that aiming for perfect knowledge is pointless.

So do it quickly, take some action, and then come back and think some more. Each time you do this, your clarity will improve, and you'll find that your understanding will grow very quickly.

Key Points

- Looking for opportunities to attach to other agendas can accelerate progress rapidly.
- Unconnected stakeholders can quickly become very supportive when they see the benefits.
- Always seek corroboration for any assumptions you are making.
- Don't forget to seek input from others.

Suggested Actions

- ○ If you haven't done so already, apply the process to the most important goal you need to be working on now.

CHAPTER 16

Building Your Reputation

The ideas and processes I've covered in this book are great at helping you to become more influential as a project manager because of the results you get and the way you build relationships. It will set you apart and provide you with a launch pad for becoming a remarkable project manager, one of high repute.

This chapter will help you to:

- Understand how to capitalise on your performance as a project manager.

- Learn how to move ahead of the pack and distinguish yourself within your organisation and profession.

- Become a "go to" person about projects in your organisation, if that's what you want.

- Create a broad base of powerful connections which will support you.

- Leverage greater impact through the power of your team.

- Become exceptional and remarkable as a project manager.

If you're in a hurry:

- You don't have to read this at all, if you are happy with where you are and what you have. As the world moves forward, the bar keeps rising and implementing what I have here will ensure you are the one raising it.

If you have a reputation for being a tough negotiator, you will influence the other party before you even meet them. Your reputation will precede you. They will be expecting a hard time; they will be working harder on their strategy and even harder on their alternatives. They may even be expecting to get a poor deal, and this may become a self-fulfilling prophecy.

Alternatively, you may have a reputation for being highly collaborative. People will know what to expect when they see you being assigned to the project. They will come to the first meeting with a positive mind, knowing that they will be listened to. You will have influenced people before you've even started, just by your reputation.

Reputations come in all shapes and sizes. What they have in common is that they are remarkable, and the owners stand out from the crowd. Influential project managers who have strong reputations for being pragmatic, flexible, tenacious, driven, opportunistic or challenging, can do far more than those who don't stand out. The processes in this book will help you to get results, and on that basis, you will have substance, but will that be enough to create a strong reputation for you?

Before I leave you, I'd like to share some ideas that can help take you to another level entirely. As with all of my ideas, they are simple — they will also help to ensure that you and your organisation benefit a great deal more from your time together.

How to Build a Reputation

If you want to become remarkable, and build a reputation that will escalate your results even further, there are six things you need to be working on:

Culture

Make sure you know the culture of the arena you want to establish and build your reputation in. The arena could be your company, profession, market or society. This awareness will help you see what will work and what won't. Unless the reputation you build will be respected and valued, you are probably wasting your time.

As you develop greater political understanding of your arena, you will start to notice what works. Go further by thinking about others who have built reputations that work for them. They may be in different roles or divisions. Seek out the people who have successfully stood out and study them. You don't need to mimic exactly what they do. See what could work for you.

Without doubt, you need to work out what the powerful people in your arena value and need in the people working there.

Distinction

For a reputation to be really useful, it has to be distinctive in the arena where it has to do its work. Don't settle for being boring and professional; find a way to really stand out from the crowd. This also means you need to be unambiguous in what you want others to expect of you.

This is more than just doing a good job. You need to think about what will distinguish you from all the other project managers. It may be that you need to become known for your tenacity, challenge, collaboration, or many other characteristics. What qualities do you think the powers that be are desperate to see in their project managers? Whatever you decide upon needs to be feasible for you to deliver. Having a little stretch is one thing, but don't try to fake it until you make it.

If you are looking to build your reputation in your profession rather than just an organisation, make sure to challenge yourself to discover what will help you to become one of the leading figures in your profession.

Substance

There is always a gap between perception and reality, but you need to make sure that it is small. If you cannot live up to your reputation, you are heading for a fall. Always make sure to focus on the performance that backs up your reputational aspirations.

As a project manager, this goes beyond being highly skilled at the practice of managing projects or programmes. Try to identify the things that other managers are not doing or displaying, in terms of capabilities. Add these to your portfolio. Get really good at demonstrating that you are using these, that you have the superior skills that back up the stunning track record you are building. Make sure that you are so good that others begin to see you as the role-model.

Don't forget that this also needs to include your chosen personal characteristics. If you have decided to be relentless and tenacious, make sure that you are demonstrating these every day, day in, day out.

Visibility

You have to ensure that people notice you. Let people see you perform and keep your focus on the attributes and behaviours that will maximise the chance that they will recognise what you want them to.

When I say people, I don't mean those you work closely with, I mean the people who rarely even get close to projects. People in other divisions and functions. Users and clients. The more people who can observe you in action, or learn from your example, the more your reputation will grow. In most organisations, project managers are only visible when they have to be. It doesn't need to be that way, and the most influential project managers will be visible to all.

This means, if you really take this point seriously, you will grow your reputation in the wider organisation. This will give you access to far more opportunities than your colleagues and put you into positions where you can influence the way everyone approaches projects. To be honest, it will help you to make sure that they don't cause problems in the future by ignoring issues until a friendly project manager has to raise it.

And I don't mean for you to become a show-off or swanky sales type. Aim to be straightforward in your approach, have a solid foundation of substance and a clear focus on what makes you distinctive.

Connections

Reputations don't work very well in isolation; they need a network of supporters, fans and advocates to spread the word. Getting others to sing your praises is far more effective than you trying to do it yourself. It is also good to associate with others who have strong reputations that complement your own.

Yes, this means networking beyond your project. Really getting to know other prominent people throughout the organisation. This can be part of your intelligence gathering activity but remember also to put on a little charm and build real relationships too. Most project managers just sit with their plans and only come out for meetings or when it is absolutely essential. Becoming the project manager that everyone knows, and likes, will be very much to your advantage. If you are diligent about displaying your substance, and have a clear idea about your distinction, you'll stand out for all the right reasons.

In time, this is the sort of activity that can easily make you a "go to" person about all things projects. Soon, people will be seeking you out as a sounding board about all of the projects they are stakeholders for. Then you are becoming really influential.

Tenacity

Consistent and relentless focus on the performance that leads to the desired reputation is essential. Trying to be one thing today, and another thing tomorrow, will confuse people and probably make them think you are unreliable (not a good reputation to have).

So, decide now and then year by year continue to climb. As you maintain this focus, the benefits will rise exponentially.

Project Teams

You don't have to build a reputation alone. If you are managing a team or project managers, make all of the ideas above a team activity. The more people you have out there promoting what project management is all about, and how people can gain the maximum benefit from it, the better your results will become.

To strengthen your team's reputation (and by implication, your own as their leader), invest some time in the following areas:

Clarify Team Purpose

It doesn't take long for the objectives to become confused. Pause a moment and clarify exactly what it is that your team is there to achieve.

Build Vision

Extending on the last point, what is your vision of project management? How should the organisation view the discipline of project management?

Consider Self-Perception

What impression do you think you and your team are creating among other groups or teams in the organisation? Are you making things difficult for others, or really helping them to succeed?

Investigate Brand Awareness

Go on, ask them. Reach out to the people you interact with and ask them in a neutral way. Make sure you motivate them to engage with your learning and ensure you don't react defensively — this part of the process should be 100% listening and learning.

Analyse Findings

Draw together your purpose, self-perception and what others think of your team. What conclusions can you draw? What gaps are there?

Translate into Benefits

For each major client/customer/work group your team interacts with, convert your team purpose into their language — what does your success help them to achieve?

Optimise Brand Values

Based on the translation of your purpose, what brand values/behaviours would maximise your ability to deliver? Also, do a reality check. Is it feasible that your team will be able to live up to these values?

Gain Team Buy-in

Because your team is so involved in the delivery of your service, they have to be totally bought-in to the impression you want them to create. The best way to do this is to involve them throughout this process and to make them part of the major decisions.

Define Behaviours

Aside from the necessary behaviours, try to capture how these are achieved/conveyed (enthusiasm, pragmatism, empathy, friendliness, etc.). The more accurately you can define behaviours, the easier it is for your team to know what is expected of them and the easier it is to see them doing it.

Consider Collateral

You might consider buying everyone a branded team T-shirt, although that may be a little extreme. Instead, think of images, documents, signatures and presentations that can be used consistently to reinforce the team brand.

Determine Measurement

If you've defined the behaviours well, you will already know what you need to measure, but when and how are you going to do this? Unless you establish a mechanism for monitoring your team's progress it will not get done and your efforts to develop your team brand will be lost within a month or two.

This is not just about simply completing the projects you and your team have been assigned to. This is about how you conduct yourselves as project managers. What you do and how you do it. The manner in which you and your team engage with people can have a dramatic effect on the way others perceive you all and how easy it is for you to get things done. Becoming exceptional project managers who are capable of rising to the challenges presented today and being ready for tomorrow. This is what influential project managers do.

Leading and Managing People

Aside from the ideas above, you can also institute the key processes explored in this book as part of the way you manage and lead people, be they team members, or part of your virtual project team. Apart from leveraging greater benefit from these processes, you will also be helping them to learn more about how to get results through influence. There are two key ways you can do this.

Performance Appraisals

If the process works for you, get others on your team to use it too. If you have people who report to you, introduce them to the process, show them how to use it and then insist that you will use it as a way of working with them.

Once they know how to do it, when they come to you for help with something they are working on, ask them to show you their Stakeholder Influence Map. Looking at the map, you can then discuss what they need to influence to achieve their result, which people are in agreement, which they need to build stronger relationships with, etc. It may not be an appropriate technique all of the time, but if they are working on important projects, and they have to work with other people around the business, it will help structure your coaching support.

And, in the process of helping them, it will also reinforce the process and its usefulness in your own work too. This leads nicely to the final idea.

Team Process

Some of the biggest benefits I have seen are where a team get together around a table to discuss what they need to make happen (influence) to achieve their goals. Then, at the right moment, the table gets split up into the different boxes of the Stakeholder Influence Map, and the discussion turns to the people who can help or hinder, and the relationship the team has with them. This quickly encourages sharing and deep debate with a clear and positive focus towards the team's goal.

In our team workshops, we often use index cards to write the stakeholders' names on. Then one individual puts the card somewhere on the table and talks through their rationale for choosing that position on the map. Team members then start to challenge and the card moves around a bit before settling into position by common consent. After the high-impact stakeholders have been discussed and positioned, the debate can then move to what needs to change — and how to make it happen.

Using the Stakeholder Influence Process in this way builds a common understanding of the problems and issues facing the team. It helps them to understand each other's perspectives, opinions and also their inside knowledge. Often, it emerges that while the team as a whole may have a poor relationship with a stakeholder, one member may have a great relationship. This can be a catalyst for improved appreciation, breaking down barriers and helping the whole team to become more effective. The team member who has the good relationship can then be called into action on behalf of the team and sent off to do some influencing.

With more people around you familiar with the process and the benefits, the more likely it will be that when you are stuck, someone else will suggest using the Stakeholder Influence Process.

The Benefits

Success that involves the help of others is made easier if you are good at influencing. The opposite is also true. Whenever you are unsuccessful, take a look at how influential you are being. Try to get into the habit that when things are not going right, when you seem to be hitting brick walls or struggling to move forward — sit back a moment and ponder two questions:

- What do you need to influence to overcome the problem?

- What do you need to influence to succeed?

As you start thinking through the answers, you may be able to spot an opportunity for a new influencing goal that you can use as the focus for the Stakeholder Influence Process. Provided it is a goal that requires change in a number of people, has natural opponents, and will take a while to achieve — it would be a good goal to focus the process on. If so, draw out a map and get to work!

Making these two questions pop into your head at the right time comes down to habit. And habits are formed through only one thing — repetition. So find ways to make it a habit, even if this means priming others to keep reminding you.

And if you can do this, on a regular basis, it is my fervent belief that you will quickly become a highly influential project manager. A project manager who is at ease with the way the projects need to work today. Not one who is stuck in the past, lamenting the way things used to work. Instead, you'll be out there, solving problems, adjusting plans and harmonising with the needs of the organisation you've been charged with supporting. You'll become a beacon of hope and someone who shows others how it's done. A true role model for getting things done in a complex world.

In addition to the many benefits your organisation will gain from this, you will prosper too. No longer stressed and frustrated. Struggling or giving up when faced with seemingly intractable issues. Instead, you'll open up a new world of work where you can have fun, you can link up with people and get things done. The power and security that can come from a demonstrable and hard-hitting track record of achievement will fuel further adventures and results.

So please, can you now just go do it? Make it work for you and reap the rewards. Come and visit us at The Influence Blog (www.learntoinfluence.com) and make sure to keep developing your approach.

Survive, thrive, and have some fun!

RESOURCES

Further Reading

- *21 Dirty Tricks at Work*, Mike Phipps and Colin Gautrey, Capstone Wiley 2005.
- *The 48 Laws of Power*, Robert Greene, Profile Books 2000.
- *The Art of Rhetoric*, Aristotle, Penguin 1991.
- *Assertiveness at Work*, Ken and Kate Back, McGraw Hill 1982.
- *Emotional Intelligence*, Daniel Goleman, Bloomsbury 1996.
- *Influence Without Authority*, Allan R. Cohen and David L. Bradford, John Wiley & Sons Inc. 2005.
- *Influence: The Psychology of Persuasion*, Dr Robert Caildini, Harper Business, 2007.
- *Influential Leadership: A Leader's Guide to Getting Things Done*, Colin Gautrey, Kogan Page, 2014.
- *The Inner Game of Work*, Timothy W. Gallwey, Villard Books 2000.

- *Managing With Power: Politics and Influence in Organizations*, Jeffrey Pfeffer, Harvard Business School Press 1994.
- *Personal Impact: What it Takes to Make a Difference,* Amanda Vickers, Steve Bavister and Jackie Smith, Pearson, Prentice Hall, 2009
- *Political Dilemmas at Work*, Dr Gary Ranker, Colin Gautrey and Mike Phipps, John Wiley & Sons 2008.
- *Political Savvy: Systematic Approaches to Leadership Behind the Scenes*, Joel R. DeLuca, EBG Publications 1999.
- *Political Skill at Work: Impact on Work Effectiveness*, Gerald Ferris, Sherryl Davidson and Pamela Perrewé, Davies Black Publishing, 2005.
- *Power, Why Some People Have It—And Others Don't*, Jeffrey Pfeffer, Collins Business 2010.
- *Working the Shadow Side: A Guide to Positive Behind the Scenes Management*, Gerard Egan, Jossey Bass Wiley, 1994.

The Gautrey Group

We work with individuals and groups all over the world, helping them to become more influential. With greater influence, careers develop, results are delivered, and increased satisfaction is achieved.

We focus exclusively on this area. So, if you want coaching, training or consultancy on how to become more influential, you'll find our insights and methods are world class. Our thought leadership is provided by Colin Gautrey, a widely recognised expert on the practical use of power and influence in the workplace.

The Collaboration Survey

Based on our research into the successes and failures of large-scale relationships, this online feedback tool collects data about the quality of each theme that needs to be strong in any great relationship (see Chapter 11).

The Gautrey Influence Profile

This unique psychometric (referred to in Chapter 12) helps clients to understand how they prefer to influence, how others may differ, and then make decisions about how to flex their style to become more influential.

Contact: The Gautrey Group

www.gautreygroup.com

info@gautreygroup.com

the gautrey group

RNC Global Projects

RNC answers the question, "How do I make sure this initiative gets the result I want?"

How? By focusing on your desired outcome, using tools and methods to support you, and engaging people with the ability to deliver. Our whole focus is helping you to achieve your plans.

With a reputation for delivering the answer no matter what, or where, RNC does what it takes to ensure you get your planned result.

RNC has global reach, multi-industry, and multi-cultural experience. So, whether your team is all in one place or dispersed throughout the world, RNC will get the results you want. Virtual project and program delivery and delivery without authority and control are RNC specialties.

Whether your initiative is in planning, needing delivery, challenged, failing or in crisis, RNC can help. Our reputation was founded on turning around projects in trouble and then delivering them. Today, astute clients start with us at the planning stage and stay with us until the celebrations.

RNC has a 16-year track record of success, with over 1000 projects and programs successfully delivered. These include core IT system replacements, biotechnology and pharmaceutical product development and manufacturing, banking and finance systems, regulation adoption projects, films, government initiatives. In fact, most things that aren't construction. Why not construction? The corporate sector needed more help.

RNC's Diane Dromgold is considered a thought-leader in the field, and she is at pains to explain the difference between project management and project delivery. Most people do the former and defend failure; RNC does the latter and enjoys your success.

Contact: RNC Global Projects

Sydney: +61 2 9238 1990

Melbourne: + 61 3 9653 9084

USA: +1 877 755 2633

Singapore: +65 6407 1154

www.rncglobal.com

info@rncglobal.com

RNC
GLOBAL PROJECTS

About Colin Gautrey

Colin is an author, trainer and executive coach who has specialised in the field of power and influence for over ten years. He combines solid research with deep personal experience in corporate life to offer his clients critical yet simple insights into how to get results with greater influence.

Based in the UK, Colin has a wealth of experience in various disciplines including Mergers and Acquisitions, International Strategy, Information Technology, Sales and Leadership Development. His passion and enthusiasm lies in the subject of influence and in helping people use this skill with integrity.

As well as providing training for an impressive portfolio of corporate clients, he has also been used by leading business schools to help students and graduates master power and influence (e.g. Wharton, London Business School and Warwick Business School). He also works with the Institute of Directors and The Conference Board in New York.

Contact Colin

twitter.com/colingautrey

linkedin.com/colingautrey

Lightning Source UK Ltd.
Milton Keynes UK
UKOW06f1802260515

252318UK00009B/568/P